G EU

INTERNATIONAL REVIEW OF NEUROBIOLOGY

VOLUME 109

SERIES EDITORS

R. ADRON HARRIS

Waggoner Center for Alcohol and Drug Addiction Research
The University of Texas at Austin
Austin, Texas, USA

PETER JENNER

Division of Pharmacology and Therapeutics
GKT School of Biomedical Sciences
King's College, London, UK

VOLUME ONE HUNDRED AND NINE

INTERNATIONAL REVIEW OF NEUROBIOLOGY

Tissue Engineering of the Peripheral Nerve

Biomaterials and Physical Therapy

Edited by

STEFANO GEUNA

Department of Clinical and Biological Sciences
Neuroscience Institute of the Cavalieri Ottolenghi Foundation (NICO)
University of Turin, Ospedale San Luigi, Regione Gonzole 10
Orbassano (TO), Italy

ISABELLE PERROTEAU

DSCB, Universita' di Torino, Regione Gonzole 10
10043, Orbassano, Torino, Italy

PIERLUIGI TOS

Traumatology Department, CTO Hospital, Via Zuretti 29
10126, Torino, Italy

BRUNO BATTISTON

Traumatology Department, CTO Hospital, Via Zuretti 29
10126, Torino, Italy

AMSTERDAM · BOSTON · HEIDELBERG · LONDON
NEW YORK · OXFORD · PARIS · SAN DIEGO
SAN FRANCISCO · SINGAPORE · SYDNEY · TOKYO
Academic Press is an imprint of Elsevier

Academic Press is an imprint of Elsevier
32 Jamestown Road, London NW1 7BY, UK
Radarweg 29, PO Box 211, 1000 AE Amsterdam, The Netherlands
The Boulevard, Langford Lane, Kidlington, Oxford, OX5 1GB, UK
225 Wyman Street, Waltham, MA 02451, USA
525 B Street, Suite 1800, San Diego, CA 92101-4495, USA

First edition 2013

ISBN: 978-0-12-420045-6
ISSN: 0074-7742

For information on all Academic Press publications
visit our website at store.elsevier.com

Printed and bound in USA
14 15 16 17   11 10 9 8 7 6 5 4 3 2 1

Working together
to grow libraries in
developing countries

www.elsevier.com • www.bookaid.org

CONTENTS

CONTRIBUTORS

Sandra Amado
Centro Interdisciplinar para o Estudo da Performance Humana (CIPER), Faculdade de Motricidade Humana, Cruz Quebrada-Dafundo, and UIS—Unidade de Investigação em Saúde, Escola Superior de Saúde de Leiria, Instituto Politécnico de Leiria, Portugal

Paulo A.S. Armada-da-Silva
Faculdade de Motricidade Humana, Universidade de Lisboa, and Centro Interdisciplinar para o Estudo da Performance Humana (CIPER), Faculdade de Motricidade Humana, Cruz Quebrada-Dafundo, Portugal

Christina Barwig
Medovent GmbH, Mainz, Germany

Bruno Battiston
Department of Traumatology, Microsurgery Unit, C.T.O. Hospital, Città della Scienza e della Salute, Turin, Italy

Lars B. Dahlin
Department of Clinical Sciences in Malmö/Hand Surgery, Lund University, Skåne University Hospital, Malmö, Sweden

Jaume del Valle
Department of Cell Biology, Physiology and Immunology, Faculty of Medicine, Institute of Neurosciences, Universitat Autònoma de Barcelona, Centro de Investigación Biomédica en Red sobre Enfermedades Neurodegenerativas (CIBERNED), Bellaterra, Spain

Thomas Freier
Medovent GmbH, Mainz, Germany

Stefano Geuna
Department of Clinical and Biological Sciences, Neuroscience Institute of the Cavalieri Ottolenghi Foundation (NICO), University of Turin, Ospedale San Luigi, Regione Gonzole 10, Orbassano (TO), Italy

Sara Gnavi
Department of Clinical and Biological Sciences, Neuroscience Institute of the Cavalieri Ottolenghi Foundation (NICO), University of Turin, Ospedale San Luigi, Regione Gonzole 10, Orbassano (TO), Italy

Claudia Grothe
Department of Clinical and Biological Sciences, Neuroscience Institute of the Cavalieri Ottolenghi Foundation (NICO), University of Turin, Ospedale San Luigi, Regione Gonzole 10, Orbassano (TO), Italy; Hannover Medical School, Institute of Neuroanatomy, and Center for Systems Neuroscience (ZSN), Hannover, Germany

Kirsten Haastert-Talini
Department of Clinical and Biological Sciences, Neuroscience Institute of the Cavalieri Ottolenghi Foundation (NICO), University of Turin, Ospedale San Luigi, Regione Gonzole 10, Orbassano (TO), Italy

Thomas Hausner
Austrian Cluster for Tissue Regeneration and Ludwig Boltzmann Institute for Experimental and Clinical Traumatology at the Research Centre for Trauma of the Austrian Workers' Compensation Board (AUVA), Vienna; Department for Trauma Surgery and Sports Traumatology, Paracelsus Medical University, Salzburg, and Department for Surgery, State Hospital Hainburg, Hainburg, Austria

Xavier Navarro
Department of Cell Biology, Physiology and Immunology, Faculty of Medicine, Institute of Neurosciences, Universitat Autònoma de Barcelona, Centro de Investigación Biomédica en Red sobre Enfermedades Neurodegenerativas (CIBERNED), Bellaterra, Spain

Antal Nógrádi
Austrian Cluster for Tissue Regeneration and Ludwig Boltzmann Institute for Experimental and Clinical Traumatology at the Research Centre for Trauma of the Austrian Workers' Compensation Board (AUVA), Vienna, Austria, and Department of Anatomy, Histology and Embryology, Faculty of Medicine, University of Szeged, Szeged, Hungary

Cátia Pereira
Faculdade de Motricidade Humana, Universidade de Lisboa, and Centro Interdisciplinar para o Estudo da Performance Humana (CIPER), Faculdade de Motricidade Humana, Cruz Quebrada-Dafundo, Portugal

Shimon Rochkind
Faculty of Life Science, Bar-Ilan University, Ramat-Gan, and Division of Peripheral Nerve Reconstruction, Department of Neurosurgery, Tel Aviv Sourasky Medical Center, Tel Aviv University, Tel Aviv, Israel

Giulia Ronchi
Department of Clinical and Biological Sciences, Neuroscience Institute of the Cavalieri Ottolenghi Foundation (NICO) University of Turin, Orbassano, Italy

Asher Shainberg
Faculty of Life Science, Bar-Ilan University, Ramat-Gan, Israel

Pierluigi Tos
Department of Traumatology, Microsurgery Unit, C.T.O. Hospital, Città della Scienza e della Salute, Turin, Italy

António P. Veloso
Faculdade de Motricidade Humana, Universidade de Lisboa, and Centro Interdisciplinar para o Estudo da Performance Humana (CIPER), Faculdade de Motricidade Humana, Cruz Quebrada-Dafundo, Portugal

PREFACE

Consensus exists among both basic and clinical scientists that peripheral nerve repair is no longer a matter of surgical reconstruction only, but rather a matter of tissue engineering which brings together several interdisciplinary and integrated treatment strategies.

In 2009, we edited a first thematic issue of the *International Review of Neurobiology* entitled "Essays on Peripheral Nerve Repair and Regeneration" (Volume 87) that collected a number of reviews on various and broad aspects of peripheral nerve regeneration research (including also several methodological papers). Following the interest raised by that book and considering the growing scientific interest on nerve repair and regeneration, we have edited this new thematic issue of the *International Review of Neurobiology* which is intended to address more specifically some of today's hot topics on peripheral nerve's tissue engineering, namely biomaterials and physical therapy.

Tissue engineering is an emerging science that finds it roots in various and complimentary disciplines (from molecular biology and biomaterials to transplantation and reconstructive microsurgery), and in order to reflect its interdisciplinary and multitranslational spirit, this thematic issue of the *International Review of Neurobiology* brings together eight reviews which aim to cover some of the most promising innovative strategies for promoting peripheral nerve repair and regeneration that emerge from basic research in the different relevant scientific areas.

The issue is introduced by a paper focused on artificial nerve scaffolds based on chitosan, a biomimetic biomaterial derived from chitin that is raising many expectations for neural repair followed by a paper on neural interfaces and neuroprostheses.

The collection continues with four reviews that critically present five different innovative approaches for promoting peripheral nerve regeneration based on physical stimulation and one more paper focused on the importance of the issue of timing in nerve repair. Finally, the last review of the book aims to throw a light on the most promising future perspectives in nerve reconstruction.

Although the papers included in this book address topics that are more specific in comparison to those addressed in the *International Review of Neurobiology* thematic issue published in 2009, all reviews have been written avoiding excessive technical details and in order to be accessible to a broad and interdisciplinary audience. It is thus expected that this collection of papers will stimulate the interest of many interdisciplinary researchers (both with basic and clinical background) and will eventually contribute to the scientific progress in tissue engineering of the peripheral nerve as well as to its successful future applications with patients suffering from nerve injury.

Stefano Geuna
Isabelle Perroteau
Pierluigi Tos
Bruno Battiston

The Use of Chitosan-Based Scaffolds to Enhance Regeneration in the Nervous System

Sara Gnavi[*], Christina Barwig[†], Thomas Freier[†],
Kirsten Haastert-Talini[‡], Claudia Grothe[‡], Stefano Geuna[*,1]

[*]Department of Clinical and Biological Sciences, Neuroscience Institute of the Cavalieri Ottolenghi Foundation (NICO), University of Turin, Ospedale San Luigi, Regione Gonzole 10, Orbassano (TO), Italy
[†]Medovent GmbH, Mainz, Germany
[‡]Hannover Medical School, Institute of Neuroanatomy & Center for Systems Neuroscience (ZSN), Hannover, Germany
[1]Corresponding author: e-mail address: stefano.geuna@unito.it

Contents

Abstract

Various biomaterials have been proposed to build up scaffolds for promoting neural repair. Among them, chitosan, a derivative of chitin, has been raising more and more interest among basic and clinical scientists. A number of studies with neuronal and glial cell cultures have shown that this biomaterial has biomimetic properties, which make it

1

a good candidate for developing innovative devices for neural repair. Yet, *in vivo* experimental studies have shown that chitosan can be successfully used to create scaffolds that promote regeneration both in the central and in the peripheral nervous system. In this review, the relevant literature on the use of chitosan in the nervous tissue, either alone or in combination with other components, is overviewed. Altogether, the promising *in vitro* and *in vivo* experimental results make it possible to foresee that time for clinical trials with chitosan-based nerve regeneration-promoting devices is approaching quickly.

1. INTRODUCTION

Chitin and its main derivative, chitosan, are becoming increasingly relevant among the novel families of biomacromolecules because of their wide potential application in biomedicine and tissue engineering (Domard & Domard, 2002; Khor & Lim, 2003; Kumar, 2002; Singh & Ray, 2000; Suh & Matthew, 2000).

Chitin and chitosan represent a family of linear polysaccharides made up of β(1–4)-linked *N*-acetyl D-glucosamine and D-glucosamine units (Domard & Domard, 2002; Muzzarelli, 1977). Depending on the processing method used to derive the biopolymer, glucosamine units may be randomly or block distributed throughout the biopolymer chain.

Chitin is primarily obtained from the exoskeleton of arthropods, shellfish such as crabs and shrimps, cuticles of insects, and cell wall of fungi (Muzzarelli, 1977). Based on the chain organization in sheets or stacks, chitin can be classified into three crystalline isoforms: α, β, and γ. The structure of α-chitin has been investigated more extensively than that of either the β or the γ form, because it is the most common polymorphic form. Commercial chitins are usually isolated from marine crustaceans, because of large amount of waste derived from food processing. In this case, α-chitin is produced while squid pens are used to produce β-chitin (Aranaz et al., 2009). Crustacean shells consist of proteins, calcium carbonate, chitin, and contain pigments such as carotenoids. Chitin is extracted by acid treatment to dissolve the calcium carbonate followed by alkaline extraction to dissolve the proteins and by a depigmentation step to obtain a colorless product (Aranaz et al., 2009; Muzzarelli, 1977).

Chitosan, on the other hand, although occurring in some fungi (Mucoraceae), is produced industrially by cleavage of the *N*-acetyl groups of the chitin *N*-acetyl D-glucosamine residues (Muzzarelli, 1977).

Chitosan is prepared by alkaline hydrolysis of acetamide groups of chitin. High temperature (100 °C) combined with strong aqueous alkali treatments are used to deacetylate chitin (N-acetylation degree lower than 30%), in order to obtain chitosan. Two different methods of preparing chitosan from chitin with varying degrees of acetylation are known: heterogeneous deacetylation of solid chitin and homogeneous deacetylation of preswollen chitin under vacuum in an aqueous medium (Aranaz et al., 2009; Muzzarelli, 1977).

The main limitations in the use of chitosan in several applications are its high viscosity and low solubility at neutral pH. Different experimental variables should be taken into account when working with chitosan solutions such as the nature of the salt counterion, length of polymer chain, molecular weight (Mw), pH, ionic strength, the addition of a nonaqueous solvent, and the degree of N-acetylation (Aranaz et al., 2009; Muzzarelli, 1977). Their different solubilities in dilute acids are commonly used to distinguish between chitin and chitosan. Chitosan, the soluble form, can have a degree of acetylation between 0% and about 60%, the upper limit depending on parameters such as processing conditions, molar mass, and solvent characteristics (Aiba, 1992). Thanks to the protonation of free amine groups present along the chitosan chain; this macromolecule can be dissolved in diluted aqueous acidic solvents, rendering the corresponding chitosan salt in solution. Degradation rate can be tuned based on its degree of deacetylation (DD), whereas fully deacetylated (DD=100%) chitosan is nondegradable (Freier, Koh, Kazazian, & Shoichet, 2005; Tomihata & Ikada, 1997) and partially deacetylated (DD=70%) chitosan is fully degradable (Tomihata & Ikada, 1997).

Chitin and chitosan are interesting candidates for use in the medical and pharmaceutical applications because they have positive properties such as biocompatibility, biodegradability, and nontoxicity that make them suitable in biomedical field (Khor & Lim, 2003). Moreover, other properties such as analgesic effect, antitumor activity, hemostatic, anticholesterolemic, antimicrobial, permeation enhancing effect, and antioxidant properties have also been reported (Kumar, Muzzarelli, Muzzarelli, Sashiwa, & Domb, 2004).

Several chitosan products have been approved by the Food and Drug Administration. Furthermore, chitosan has been used to generate laser-activated film surgical adhesive (SurgiLux) that can be very useful as an alternative to microsurgery for peripheral nerve reconstruction (Foster & Karsten, 2012). SurgiLux has been tested *in vitro* and *in vivo* on different tissues including nerve, intestine, *dura mater*, and cornea, demonstrating a good biocompatibility (Foster & Karsten, 2012).

The American Society of Testing Materials (ASTM F04 division IV) is making efforts to establish standard guidelines for tissue engineered medical products (ASTM, 2001). The F2103 guide covers the evaluation of chitosan salts suitable for use in medical applications. Moreover, *chitosan hydrochloride* (a derivative of chitosan) has been included in the European Pharmacopeia in 2002 (Pharmacopeia, 2002).

In order to be approved as a biomedical material, sterility is an important issue to resolve. Chitosan products intended for parenteral administration and those in contact with serous fluids, for example, wounds, have to be sterilized before use. Common methods for the sterilization include exposure to dry heat, saturated steam, ethylene oxide, or γ radiation. Before using any of these methods for chitosan product sterilization, their effects on polymer properties and end performance have to be tested, as they can cause irreversible damage to the morphological, physical, mechanical, and biological characteristics. Dry heat sterilization method resulted in lower aqueous solubility for chitosan and in insolubility in acidic aqueous media (Lim, Khor, & Ling, 1999). Saturated steam and γ irradiation caused an acceleration in the rate and extent of chitosan chain scission events, respectively. The use of 70% ethanol, as a sterilizing agent, is a suitable method as it did not alter chitosan–membrane characteristic; however, it is limited to small-scale applications. Ethylene oxide is a simple technique to be used for sterilization of industrially produced chitosan membrane, preserving chitosan–membrane morphology, percentage of strain at break, and *in vitro* cytotoxicity to Vero cells. Moreover, this method can be used for industrial sterilization of chitosan membrane (Marreco, da Luz Moreira, Genari, & Moraes, 2004). The long-term storage may have effects and implications on the integrity of chitin and chitosan materials (Kam, Khor, & Lim, 1999), and further investigation is needed to optimize sterilization and storage method conditions.

It has been shown that chitosan is capable of forming large phospholipid aggregates by inducing the fusion of small dipalmitoyl phosphatidylcholine bilayers, which are a major component of the plasma membrane (Pavinatto et al., 2007; Zuo et al., 2006). Thus, the use of chitosan as a "fusogen" might be more advantageous as a potential clinical tool relative to nonionic polymers (e.g., PEG or P188).

Due to its high biodegradability and biocompatibility, together with its specific interactions with components of the extracellular matrix (ECM) and growth factors, chitosan employment is growing in a variety of applications, including implantable and injectable orthopedic and periodontal systems, drug delivery systems, wound-healing agents, lung surfactant additives, and tissue engineering scaffolds for tissue regeneration of skin, bone, and cartilage (Drury & Mooney, 2003; Gan & Wang, 2007; Janes, Fresneau, Marazuela, Fabra, & Alonso, 2001; Madihally & Matthew, 1999; Roy, Mao, Huang, & Leong, 1999; Suh & Matthew, 2000; Ueno et al., 1999; Yuan, Zhang, Yang, Wang, & Gu, 2004; Zuo et al., 2006). Yet, chitosan is a versatile material currently used in clinical wound dressings, primarily for its hemostatic property (Gustafson, Fulkerson, Bildfell, Aguilera, & Hazzard, 2007).

Another important application of chitosan is the development of drug delivery systems such as nanoparticles, hydrogels, microspheres, films, and tablets. The abundance of primary amine groups enables chitosan to be ionically or covalently coupled to various biomolecules because the amine moieties become predominantly protonated and positively charged below pH 6.5, whereas they are increasingly deprotonated at pH 6.5 and above. As a result of its cationic character, chitosan is able to react with polyanion giving rise to polyelectrolyte complexes (Acosta, Aranaz, Peniche, & Heras, 2003; Peniche, Arguelles-Monal, Peniche, & Acosta, 2003).

Moreover, due to its positive charge, chitosan can interact with negative molecules such as DNA. This property has been used to prepare a nonviral vector gene delivery system (Mumper, Wang, Claspell, & Rolland, 1995).

Among the different tissue organs, many studies have investigated the use of chitosan for repair, not only because of its biocompatibility, biodegradability, low toxicity, and cost but also because of its excellent potential for supporting three-dimensional organization of regenerating tissues (Evans et al., 1999; Ho et al., 2005; Ma et al., 2003; Madihally & Matthew, 1999; Novikova, Novikov, & Kellerth, 2003; Vasconcelos & Gay-Escoda, 2000).

Here, we review the main chitosan-based bioengineering strategies for peripheral nerve and spinal cord injury (SCI) repair. This review has been divided into the following sections: in the first part, we report *in vitro* studies on the evaluation of chitosan properties, the second and the third parts cover *in vivo* studies for spinal cord and peripheral nerve repairs, respectively.

2. *IN VITRO* EVIDENCE: CHITOSAN PROPERTIES, BIOCOMPATIBILITY, AND SURFACE MODIFICATION

2.1. Chitosan physical properties

2.1.1 Mechanical strength

Chitosan matrices have been shown to have low mechanical strength under physiological conditions (Itoh et al., 2003; Madihally & Matthew, 1999) and to be unable to maintain a predefined shape for transplantation, which has initially limited their use as nerve guidance conduits (NGC; Freier, Montenegro, Koh, & Shoichet, 2005; Itoh, Suzuki, et al., 2003; Itoh et al., 2003; Yamaguchi, Itoh, Suzuki, Osaka, & Tanaka, 2003).

To improve chitosan mechanical properties, Ao et al. (2006) used a novel mold and thermally induced phase-separation technique with a unidirectional temperature gradient to produce chitosan conduits containing longitudinally aligned microfibers. This preparation method may allow the incorporation of therapeutic agents into the matrix for sustained release, as no toxic substances have been used (Ao et al., 2006). *In vitro* characterization using Neuro-2a-cells verified that the mold-based multimicrotubule chitosan conduit had suitable mechanical strength, microtubule diameter distribution, porosity, swelling, biodegradability, and nerve cell affinity, for applications in nerve tissue engineering (Ao et al., 2006).

A novel method to create porous tubular chitosan scaffolds with desirable mechanical proprieties and controllable inner structure has been developed by Wang et al. (2006). Inner matrix, with multiple axially oriented macrochannels and radially interconnected micropores, was produced using acupuncture needles as mandrel during the molding process (Wang et al., 2006). *In vitro* characterization demonstrated that the scaffolds possessed suitable mechanical (porosity, swelling, and biodegradability) and biological (differentiated Neuro-2a cells grew along the oriented macrochannels) properties for applications in nerve tissue engineering (Wang et al., 2006). However, these scaffolds have a low mechanical strength under physiological conditions, thus limiting their applicability. In order to increase mechanical strength, chitosan conduits have been reinforced with additives (Yang et al., 2004) or cross-linked with chemical substances such as formaldehyde (Wang et al., 2005).

Recently, a mold-casting/lyophilization method was used to fabricate porous chitosan nerve conduits; however, these conduits still have a low mechanical strength under physiological conditions. The porous structure of the chitosan conduit was reinforced by introducing braided chitosan fibers, leading to an increase in tensile strength (Wang et al., 2007). These conduits were permeable to molecules ranging in molecular size from 180 to 66,200 Da (Wang et al., 2007). Moreover, *in vitro* direct contact cytotoxicity test, using Neuro-2a cells, showed that the conduits were not cytotoxic (Wang et al., 2007).

While chitosan has low mechanical strength under physiological conditions (Madihally & Matthew, 1999; Yamaguchi et al., 2003; Yang et al., 2002), chitin gels prepared by selective N-acetylation of chitosan amine groups (Hirano, Ohe, & Ono, 1976) are known to be mechanically strong (Vachoud & Domard, 2001), suggesting that they may be able to overcome the insufficient strength described for chitosan-based NGCs. Chitosan hydrogel tubes have been fabricated from chitosan solution using acylation chemistry and mold-casting techniques followed by alkaline hydrolysis that results in chitosan tube formation, with the extent of hydrolysis controlling the resulting amine content (Freier, Montenegro, et al., 2005). Chitosan tubes resulted to be mechanically stronger to support adhesion and differentiation of primary chick dorsal root ganglion neurons and to significantly enhance neurite outgrowth (Freier, Montenegro, et al., 2005).

Also, the DD affects chitosan mechanical properties, and it has been shown that the swelling index of chitosan films decreases and the elastic modulus and tensile strength increase with the increase in DD (Wenling et al., 2005).

Finally, Wang et al. showed that a bilayered chitosan mesh tube, with an inner layer of oriented nanofibers and an outer layer of random nanofibers, increased the resistance to the compression force compared with the random fiber mesh tubes (Wang, Itoh, Matsuda, Ichinose, 2008; Wang et al., 2008).

2.1.2 Porosity

Porosity of a regenerative scaffold is an important factor in tissue engineering. Huang et al. described an easy method for the production of longitudinally oriented channels made of chitosan by using a lyophilizing and wire-heating process. Ni–Cu wires were used as a mandrel because of their high level of resistance (Huang, Huang, Huang, & Liu, 2005). In comparison with poly-lactic-*co*-glycolic acid (PLGA), the permeability and porous structure of chitosan improved its effectiveness for the nerve tissue engineering (Huang, Onyeri, Siewe, Moshfeghian, & Madihally, 2005). The employ of a weak base, to neutralize chitosan, can be used to influence the porous structure, making it more uniform (Huang, Onyeri, et al., 2005). Porosity geometry can also be controlled by a production method based on inverted colloid crystals (Kuo & Lin, 2013).

2.1.3 Chitosan biodegradability

Concerning biodegradability, chitin and chitosan are degraded *in vivo* by proteases present in all mammal tissues, such as lysozyme, papain, and pepsin, leading to the release of nontoxic oligosaccharides of variable length which can be incorporated into glycosaminoglycans and glycoproteins (Pangburn, Trescony, & Heller, 1982). The length of the chains also affects the degradation rate (Huang, Khor, & Lim, 2004; Tomihata & Ikada, 1997; Zhang & Neau, 2001). Controlling degradation rate of chitin- and chitosan-based biopolymers is essential in drug delivery and tissue regeneration applications. The degradation rate also affects the biocompatibility as fast degradation rate results in amino sugars accumulation that may lead to inflammatory response. Chitosan samples with low DD may induce an inflammatory response, whereas chitosan samples with high DD do not because of the low degradation rate (Hirano, Tsuchida, & Nagao, 1989; Kurita, Kaji, Mori, & Nishiyama, 2000; Sashiwa, Saimoto, Shigemata, Ogawa, & Tokura, 1991).

The degradation rate of chitosan can be influenced by physical parameters such as porosity, fiber diameter, blending with other polymers, or the use of cross-linking agents. Chitosan scaffolds with high porosity degrade faster than scaffolds with smaller pore diameter. Within the same range of porosity, scaffolds with smaller pore diameter degrade faster (Cunha-Reis et al., 2007).

Adjusting the pH of the solution of a nerve conduit has been reported to influence degradation properties. In particular, increasing the layer numbers and overcoming the acidity-caused autoacceleration of poly-D, L-lactic acid/chondroitin sulfate/chitosan (PDLLA/CS/CHS) nerve conduit decrease its biodegradability rate retaining its integrity up to 3 months (Xu, Yan, Wan, & Li, 2009).

The degradation kinetics is inversely related to the crystallinity degree which can be controlled by acting on the DD and on the distribution of acetyl groups. The absence of acetyl groups or their homogeneous/random distribution results in low enzymatic degradation rates (Aiba, 1992; Suh & Matthew, 2000). Chitosan nano-/microfiber meshes with a deacetylation of 78% have a faster biodegradation rate than meshes with a deacetylation of 93% and collapse over the time, causing occlusion of the tube made from these meshes (Wang, Itoh, Matsuda, Ichinose, et al., 2008). Degradation of chitosan films with very low (about 0.5%) or high (about 99.2%) acetylation is minimal over 4-week period, whereas progressive mass loss to greater than 50% has been reported for chitosan film with 30–70% acetylation (Freier, Koh, et al., 2005).

Blending of chitin with other biomaterials like gelatin results in a faster degradation rate and significant loss of material compared with chitosan alone (Huang, Onyeri, et al., 2005).

Finally, the use of cross-linking agents such as hexamethylene diisocyanate (HDI), epichlorohydrin (ECH), and glutaraldehyde (GA) may also result in significant decrease in degradation rate compared with noncross-linked chitosan (Cao et al., 2005).

2.2. Chitosan biocompatibility

Biocompatibility properties of chitin and chitosan depend on the sample characteristics such as natural source, preparation method, Mw, and DD. In particular, residual proteins, in chitin and chitosan, derived from production methods could cause allergic reactions such as hypersensitivity. Biocompatibility is an important issue in the choice of a biomaterial for peripheral nerve regeneration or SCI repair.

In vitro studies have shown that chitosan exerts biomimetic in the peripheral nervous system, allowing neuronal adhesion, differentiation, and growth (Cheng, Cao, et al., 2003; Yang et al., 2004). Different blends of chitosan and gelatin cross-linked with genipin allow cell adhesion and proliferation of NIH3T3 mouse fibroblasts and S5Y5 neuroblastoma cells. Cross-linked samples were found to be biocompatible in particular blends containing 8% gelatin supporting very well neuroblastoma cell adhesion and proliferation (Chiono et al., 2008). Among different blends of chitosan with polyacrylamide, ethyl acrylate, and hydroxyethyl acrylate, chitosan, poly(methyl acrylate), and 50% (w/w) blends of ethyl acrylate and hydroxyethyl acrylate were the most suitable polymers to promote *in vitro* cell adhesion and differentiation of neural explants from the medial ganglionic eminence and the cortical ventricular zone of embryonic rat brains (Soria et al., 2006).

Polymeric biomaterial composite of chitin, chitosan, and gelatin, with a pore geometry of inverted colloidal crystals, induced pluripotent stem (iPS) cell adhesion and proliferation and has the potential to guide and accelerate differentiation of iPS cells toward a neuron phenotype (Kuo & Lin, 2013).

It has been also shown that rat pheochromocytoma cell line (PC12), grown on chitosan–gelatin–fibronectin-composed films, differentiates more rapidly and extends longer neurite than on pure chitosan films (Cheng, Deng, et al., 2003).

Another study reported that the blending of chitosan with polycaprolactone (PCL) increased cell viability and redistribution of actin cytoskeletal fibers of mouse embryonic fibroblasts cultured *in vitro* (Sarasam & Madihally, 2005).

In addition, biodegradable chitosan microgrooved polymers were successfully used to align Schwann cells (SCs) and cells of the glial cell line C6. SCs display high degree of alignment and express neurotrophic factors, like glial-derived neurotrophic factor (GDNF) and nerve growth factor (NGF; Hsu, Lu, Ni, & Su, 2007). Yuan et al. also reported that chitosan membranes and fibers have excellent neuroglial cell affinity and good biological compatibility. SCs grown on chitosan membranes displayed a spherical shape, whereas on chitosan fibers, they had an elongated morphology (Yuan et al., 2004). Yet, Wang et al. (2009) reported that a chitosan non-woven nanofiber mesh tube with an inner layer of oriented nanofibers, produced by electrospinning method, induced alignment of immortalized adult mouse SCs (IMS32).

Cells of the rat SCs line RT4-D6P2T cultured on PCL/chitosan blend nanofibrous scaffold showed higher cell proliferation in comparison with cells grown on PCL scaffolds alone and maintained their characteristic cell morphology, with spreading bipolar elongation (Prabhakaran et al., 2008). Interestingly, culturing adult rat SCs on chitosan films with low acetylation resulted in better cell spreading and proliferation (Wenling et al., 2005).

Cross-linking agents may also influence the adhesion and proliferation of cultured SCs. Cao et al. (2005) showed that HDI cross-linked chitosan films enhanced the spread and the proliferation of SCs, whereas ECH and GA cross-linked films delayed cell proliferation.

Electrical stimulation, through conductive polymers, can be used to enhance neurite outgrowth and peripheral nerve regeneration. Conductive polypyrrole chitosan membranes have been shown to support SC adhesion, spreading with and without electrical stimulation (Huang, Hu, et al., 2010). Interestingly, this study provides confirmation of cell biocompatibility on chitosan conductive polymers.

Concerning employability of chitosan for neural repair, *in vitro* studies using neural stem cells (NSCs), human adipose-derived stem cells (hADSCs), neuroepithelial stem cells (NEPs), and iPS cells have been carried out since these cells have a great potential as a cell replacement therapy for SCI.

Chitosan/collagen membrane showed low cytotoxicity supporting rat NSC (at the neurosphere level) survival, proliferation, and differentiation. In particular, cells migrate out from the neurospheres and differentiate into neurons (Yang, Mo, Duan, & Li, 2010). NSCs have been also cultured on laminin-coated chitosan channels (Guo et al., 2012).

Moreover, chitosan carrier loaded with neurotrophin 3 (NT-3) pro-vided an ideal environment for adhesion, proliferation, and differentiation of NSCs (Yang, Duan, Mo, Qiao, & Li, 2010).

NSCs seeded in fibrin scaffolds within a chitosan channel, containing PLGA microspheres releasing dibutyryl cyclic-AMP, differentiate *in vitro* to β-III-tubulin positive neurons providing further confirmation of chitosan scaffold biocompatibility (Kim, Zahir, Tator, & Shoichet, 2011).

Yet, scaffolds made of chitin, chitosan, and gelatin with pore geometry of inverted colloidal crystals have been successfully used to guide the differen-tiation of iPS cells into neurons (Kuo & Lin, 2013).

Finally, hADSCs have been successfully transdifferentiated from mesen-chymal into the neural lineage onto a chitosan-coated surface (Hsueh, Chiang, Wu, & Lin, 2012). hADSCs are a subset of multipotent mesenchy-mal stem cells with less ethical conflict and minimal invasive surgical proce-dure to obtain cells and can thus be tentatively proposed as agents for promoting nerve regeneration in patients.

NEPs are NSCs with multipotentiality for neuronal and glial differentia-tion. NEPs have been reported to adhere and grow on chitosan fibers. More-over, they could differentiate into neurons and glia (Fang et al., 2010). This study demonstrated that chitosan fibers have good biocompatibility with NEPs.

2.3. Chitosan surface modification

Improving nerve cell affinity for chitosan is a key issue for improving its effectiveness for neural regeneration (Dhiman, Ray, & Panda, 2004; Haipeng et al., 2000; Zhu, Gao, He, Liu, & Shen, 2003; Zielinski & Aebischer, 1994). Combining chitosan with poly-L-lysine, laminin, laminin peptide, or collagen may increase cell adhesion, growth, and viability.

Blending chitosan with poly-L-lysine improved PC12 cell affinity in comparison with chitosan and chitosan–collagen films as demonstrated by increasing attachment, growth, and differentiation into nerve cells. The increased cell affinity might be due to both the increased surface charge and hydrophilicity of composite materials (Mingyu et al., 2004).

Another study also reported that poly-L-lysine-, collagen-, or albumin-blended chitosan exhibit better nerve cell affinity, neurite outgrowth, and proliferation of PC12 and fetal mouse celebral cortex cells than original chitosan (Cheng, Cao, et al., 2003).

Thermoresponsive chitosan/glycerophosphate salt hydrogel coated with poly-D-lysine immobilized via azidoaniline photocoupling improves cell adhesion and morphology and neurite outgrowth compared with uncoated chitosan/glycerophosphate salt hydrogel (Crompton et al., 2007). Increasing poly-D-lysine concentration did not alter cell survival but significantly inhibited neurite outgrowth (Crompton et al., 2007).

Laminin is an 180-KDa glycoprotein that plays an important role in neuronal cell adhesion, differentiation, and neurite outgrowth (Madison, da Silva, Dikkes, Sidman, & Chiu, 1987; Manthorpe et al., 1983). Two peptides of the lamin-1 molecule, namely, YIGSR (Tyr-Ile-Gly-Ser-Arg) and IKVAV (Ile-Lys-Val-Ala-Val) sequences, mediate receptor-specific neural cell adhesion and are known to promote cell adhesion and neurite outgrowth, respectively (Graf et al., 1987; Kleinman et al., 1988; Pierschbacher & Ruoslahti, 1984; Sephel, Burrous, & Kleinman, 1989; Tashiro et al., 1989). Moreover, these domains enhance SC migration. Surface modification of a biomaterial may improve its biocompatibility. Matsuda et al. developed a new biomaterial consisting of molecularly aligned chitosan with IKVAV and YIGSR peptides bonded covalently. Briefly, chitosan was thiolated by reacting 4-thiobutyrolactone with the chitosan amino group and thiol group of cysteine located at the end of the synthetic laminin peptides that were reacted chemically with thiolated chitosan to form chitosan-S-S-laminin peptide (Matsuda, Kobayashi, Itoh, Kataoka, & Tanaka, 2005).

A novel chitosan gel has been synthesized by reaction of chitosan amine group with methacrylic anhydride, resulting in methacrylamide chitosan (Yu, Kazazian, & Shoichet, 2007). Maleimide-terminated cell adhesive peptides, mi-GDPGYIGSR and mi-GQASSIKVAV, have been coupled to a thiolated form of methacrylamide chitosan, resulting in increased neuronal adhesion and neurite outgrowth (Yu et al., 2007).

Synthetic surface modification methods often lead to alterations of the original material's physical proprieties. The plasma surface modification process has been shown to be able to modify the surface properties of a biomaterial without affecting its bulk physical properties (Yeh, Iriyama, Matsuzawa, Hanson, & Yasuda, 1988). Compared with the conventional chemical method, the percentage of laminin incorporated on chitosan films by plasma treatment is significantly higher (Huang, Huang, Huang, & Chen, 2007). Moreover, laminin-modified chitosan membrane significantly increases SC adhesion (Huang et al., 2007). Carbon nanotube/chitosan fibers coated with laminin, via an oxygen plasma technique, allowed PC12 cell adhesion, growth, and guided oriented neurite outgrowth (Huang et al., 2011).

Gliosarcoma cells (9 L) and primary neurons have been cultured on chitosan, GA-cross-linked chitosan, GA-cross-linked chitosan–gelatin conjugate, a chitosan–gelatin mixture, chitosan coated with poly-L-lysine, chitosan coated with laminin, and chitosan coated with serum revealed that coated chitosan, especially chitosan coated with laminin, has excellent nerve cell affinity, promoted better cell adhesion, spread, and growth in comparison with cross-linked-chitosan or chitosan alone (Haipeng et al., 2000).

In conclusion, it was shown that chitosan precoated with ECM molecules, in particular laminin, improves nerve cell affinity. The ECM molecules adsorbed on the materials, and the physicochemical properties of the material improve the adhesion and spread of nerve cells on the biomaterials.

2.4. Chitosan as a tool for neurotrophic factor delivery

To enhance axonal regeneration, spatial and temporal delivery of therapeutic molecules combined with biomaterials may be helpful. Current methods for therapeutic agent delivery, such as oral and intravenous administration, are inadequate for local delivery as they have limitation of dose control, premature drug degradation, not specific action, and may lead to undesirable side-effects and/or system effects.

In order to provide a system for local and sustained growth factor release, poly-lactide-*co*-glycide microspheres have been incorporated into chitosan guidance channels by spin-coating the interior of a chitosan channel with a chitosan solution containing microspheres minimizing the exposure of PLGA microspheres to acidic solution (Kim, Tator, & Shoichet, 2008). Alkaline phosphatase, used as a model protein to test the release and bioactivity, showed high encapsulation efficiency and bioactivity profile over a 90-day period *in vitro* (Kim et al., 2008).

Poly-lactide-*co*-glycide microspheres have been physically entrapped in between two concentric tubes consisting of a chitosan inner tube and a chitin outer tube (Goraltchouk, Scanga, Morshead, & Shoichet, 2006). Bovine serum albumin (BSA), used as a model drug, was released up to 84 days after encapsulation in the microspheres (Goraltchouk et al., 2006). Epidermal growth factor (EGF), coencapsulated with BSA, was released for 56 days with a similar profile to that of BSA and was found to be active up to 14 days (Goraltchouk et al., 2006).

Moreover, microstructured polymer filaments used as a nerve implant have been successfully loaded with chitosan/siRNA nanoparticles to promote nerve regeneration and ensure local delivery of nanotherapeutics. The nanoparticles were internalized by the cells resulting in target mRNA reduction and enhanced neurite outgrowth (Mittnacht et al., 2010).

Pfister et al. reported that NGF release kinetics could be regulated by embedding NGF at different radial locations within a nerve conduit. In particular, polyelectrolyte alginate/chitosan conduit was coated with poly(lactide-*co*-glycolide) to control the release of the embedded NGF (Pfister, Alther, Papaloizos, Merkle, & Gander, 2008). A sustained release of NGF in low nanogram concentration per day was obtained for up to 15 days *in vitro* (Pfister et al., 2008).

Hydrogels are cross-linked polymers characterized by high water content. They can have a possible application as growth factor-releasing systems. Biodegradable hydrogels made up of oppositely charged polysaccharide alginate and chitosan showed high water uptake (84% (w/w)) and permitted good permeation of fluorescent-labeled dextran in a molecular-weight-dependent manner (Pfister, Papaloizos, Merkle, & Gander, 2007).

Spatially defined patterns of NGF can be created using photochemical immobilization technique made possible by UV confocal laser pattering. NGF has been chemically immobilized on chitosan films in distinct areas or as concentration gradients remaining bioactive as demonstrated by *in vitro* culturing of dissociated primary neuron from rat superior cervical ganglia. When neurons are plated on a chitosan film characterized by distinct immobilized NGF-patterned areas surrounded by unmodified chitosan, remained as single spread cells in the NGF-patterned region and formed clusters resulting in lower cell survival in the unmodified chitosan areas. Moreover, the immobilized NGF induces axonal sprouting compared with the unmodified chitosan (Yu, Wosnick, & Shoichet, 2008).

Finally, a recent study showed that photo-cross-linkable streptavidin-modified methacrylamide chitosan 3D hydrogel, along with the recombinant biotin–interferon-γ promotes neuronal differentiation of neuronal stem/progenitor cells (NSPCs, Leipzig, Wylie, Kim, & Shoichet, 2011).

3. CHITOSAN FOR CENTRAL NERVOUS SYSTEM REPAIR

So far, the potential use of chitosan for central nervous system (CNS) nerve repair has been focused on SCI, a common outcome of traffic accidents like motor vehicle crashes, sports injuries, and trauma that may lead to life-long paralysis, for which, unfortunately, there is no effective cure. In fact, axons of adult mammals regenerate poorly and in a disorganized manner or fail to regenerate spontaneously after SCI. When nervous tissue loss occurs, different methods have been used to bridge the spinal cord gap. For example, transplantation of peripheral nerves (Bray, Villegas-Perez, Vidal-Sanz, & Aguayo, 1987; Cheng, Cao, & Olson, 1996), SCs (Bunge, 2002; Novikova, Pettersson, Brohlin, Wiberg, & Novikov, 2008; Xu, Zhang, Li, Aebischer, & Bunge, 1999), olfactory ensheathing cells (Li, Field, & Raisman, 1997; Ramon-Cueto, Plant, Avila, & Bunge, 1998), and NSCs has been used (Teng et al., 2002; Xue et al., 2012; Zheng & Cui, 2012). These studies have shown that CNS axons can regenerate in an appropriate microenvironment and injured axons can recover part of their function. However, the above-mentioned methods have limitations for clinical applications, such as damage to the donors of peripheral nerves and immunological rejection.

Biomaterials are becoming increasingly popular as a potential tool for the treatment of SCI as a mean to restore the ECM at the site of injury. Various materials, both of natural and synthetic origin, have been investigated for potential applications in the spinal cord (Nomura, Tator, & Shoichet, 2006; Novikova et al., 2003; Samadikuchaksaraei, 2007; Straley, Foo, & Heilshorn, 2010). These materials can support endogenous tissue regeneration (Tysseling-Mattiace et al., 2008; Woerly, Pinet, de Robertis, Van Diep, & Bousmina, 2001), promote directed axonal regrowth (Li and Hoffman-Kim, 2008; Yoshii, Ito, Shima, Taniguchi, & Akagi, 2009), enhance cell transplant survival and integration (Itosaka et al., 2009; Teng et al., 2002), deliver drugs (Johnson, Parker, & Sakiyama-Elbert, 2009; Kang, Poon, Tator, & Shoichet, 2009; Willerth & Sakiyama-Elbert, 2007), and seal damaged *dura mater* (Gazzeri et al., 2009). Biomaterials designed for spinal cord repair should provoke minimal chronic inflammation and immune responses when implanted into the body (Anderson, Rodriguez, & Chang, 2008; Williams, 2008). These responses depend not only on the intrinsic properties of the material itself but also on the form in which the material is presented, for example, implant shape (Di Vita et al., 2008), size (Kohane et al., 2006), and porosity (Ghanaati et al., 2010). In particular, it is important to monitor over time degradation kinetics and secondary product formation of biomaterials because degradation products can elicit inflammatory responses that may be different than those elicited by the implanted material. Regarding degradation kinetics, chitosan is an attractive material because of its degradation rate that can be regulated by acting on its DD. Fully deacetylated (DD=100%) chitosan is nondegradable (Freier, Koh, et al., 2005; Tomihata & Ikada, 1997), whereas partially deacetylated

(DD=70%) is fully degradable (Kofuji, Ito, Murata, & Kawashima, 2001; Tomihata & Ikada, 1997). In recent years, chitosan, either alone or in combination with other biomaterials (Table 1.1), adhesion peptides (Table 1.2), supportive cells (Table 1.3), or growth factors (Table 1.4), has been widely used for spinal cord repair.

Ex vivo and *in vivo* SCI models demonstrate that chitosan is able to restore compromised membrane integrity following spinal cord trauma, reduces injury-mediated production of reactive oxygen species (ROS), and restricts continuing lipid peroxidation, displaying a potent neuroprotective role even though it did not show any ROS, or acrolein, scavenging ability (Cho et al., 2010). Yet, the use of chitosan has therapeutic potential through site-specific delivery following traumatic spinal cord and head injury (Cho et al., 2010).

To increase the potential for axonal regeneration and functional recovery, implantation of autograft combined with biomaterials appears to be a promising strategy too. Nomura, Baladie, et al. (2008) have shown that intracavitary implantation of chitosan guidance channels containing peripheral nerve grafts after subacute SCI resulted in a thicker bridge containing a larger number of myelinated axons compared with chitosan channels alone. Peripheral nerve-filled chitosan conduits showed an excellent biocompatibility with the adjacent neural tissue with no signs of degradation and minimal tissue reaction at 14 weeks after implantation (Nomura, Baladie, et al., 2008).

3.1. Surface modification of chitosan conduits for CNS repair

A promising strategy for facilitating nerve regeneration is the combination of biomaterials with adhesion molecules (Table 1.2), such as laminin (Cheng et al., 2007; Lemmon, Burden, Payne, Elmslie, & Hlavin, 1992), L1 (Lemmon et al., 1992), N-cadherin (Lemmon et al., 1992), and collagen (Li et al., 2009). These molecules may be positioned in the inner portion of the tube device in order to guide neurite growth. Chitosan conduits enriched with adhesion molecules have been already used *in vivo* with the goal of better directing the repair of damaged axons following SCI. Biodegradable porous chitosan nerve conduits, filled with semifluid type I collagen, have been developed using lyophilizing and wire-heating process (Li et al., 2009) and implanted into the injured spinal cord of a rat model. Results showed that collagen serves as a directional guide to facilitate correctly aligned axon regrowth and enhances nerve regeneration across a gap. Yet, the chitosan tube blocked the invasion of glial scar tissue into the lesion site (Li et al., 2009).

Table 1.1 Types of chitosan conduits used in spinal cord repair

Intrinsic framework

Nerve tube	Framework	Internal filler	Animal	Injury	Gap size (mm)	Methods	Controls	Follow-up	Outcome	Authors and year
Chitosan	—	—	Female Sprague–Dawley rats	Laminectomy at T8 vertebral level	2	Histological evaluation	Degradable polyglycolide and nondegradable expanded polytetrafluoroethylene tubes	12 months	Chitosan, in comparison with the other tested materials, does not elicit immune response. At 12 months, postimplantation chitosan is not degraded	Kim et al. (2011)
Chitosan membrane	—	—	Guinea pigs	Complete transection or compression injury in the midthoracic region	—	Somatosensory evoked potential (SSEP)	Subcutaneous injection of Ringer's solution	2 weeks	Topical application of chitosan after complete transection or compression restored the conduction of nerve impulses through the length of spinal cord	Cho, Shi, and Borgens (2010)
Chitosan tube	Rat intercostal nerve from T7 to T11	Peripheral nerve graft	Female Sprague–Dawley rats	Laminectomy at T7–T9 and spinal cord compression at T8 with a 50-g clip for 1 min	8 mm length and 1.8 mm diameter	Basso–Beattie–Bresnahan (BBB) test, anterograde axonal tracing with biotin dextran amine, and histological evaluation	Empty chitosan tube	14 weeks	Chitosan tube, containing peripheral nerve graft, contains a higher number of myelinating axons compared with chitosan tube alone. SCs from the peripheral nerve graft have high myelination capacity. Chitosan tube shows excellent biocompatibility	Nomura et al. (2008)

Table 1.2 Modifications to chitosan conduits surface used in central nervous system repair

Surface modification

Nerve tube	Surface modification	Internal filler	Animal	Injury	Gap size (mm)	Methods	Controls	Follow-up	Outcome	Authors and year
Chitosan tube	Collagen type-I	Semifluid collagen type-I	Wistar rats	Laminectomy from T7 to T10 vertebrae + lateral incision at T9 to excise a segment of spinal cord of 4 mm length and 2 mm width	About two-third spinal cord	Basso–Beattie–Bresnahan (BBB) test, anterograde axonal tracing with biotin dextran amine (BDA), retrograde axonal tracing with FluoroGold (FG), and histological evaluation	Empty chitosan tube and lesion without tube implantation	12 months	Axons from the proximal spinal cord regenerate, cross the lesion area inside the tube, and lead to functional restoration of paralyzed hind limbs	Li, Yang, Zhang, Wang, and Chen (2009)
Porous chitosan tube	Laminin coating of the inner surface by oxygen plasma treatment	–	Female Sprague–Dawley rats	Complete transection at T8 and removal of 5 mm piece of spinal cord tissue	5	Basso–Beattie–Bresnahan (BBB), CCombined Behavior Score (CBS) test, and histological evaluation	Empty chitosan tube	2 months	Laminin-coated chitosan tube improves functional recovery by guiding damaged axon regrowth, through the lesioned area, without inducing inflammation or apoptosis	Cheng, Huang, Chang, and Huang (2007)
Polyglycolic acid–chitosan tube	Coating with recombinant L1-Fc	–	Rats	Optic nerve transection	–	Anterograde and retrograde tracing and histological evaluation	Polyglycolic acid–chitosan tube	–	The polyglycolic acid–chitosan conduit coated with L1-Fc is more effective to promote axonal regeneration and remyelination	Xu et al. (2004)

Table 1.3 Types of supportive cells used to enhance chitosan tubes in spinal cord repair Combination with cells

Nerve tube	Cell type	Internal filler	Animal	Injury	Gap size (mm)	Methods	Controls	Follow-up	Outcome	Authors and year
Laminin-coated chitosan tube	Neural stem/progenitor cells (NSPCs)	NSPCs preseeded on laminin-coated chitosan tube in combination with nogo–66 receptor protein, basic fibroblast growth factor (bFGF), epidermal growth factor (EGF), and platelet–derived growth fator (PDGF)	Female Sprague–Dawley rats	Complete spinal cord transection T7–T9 laminae were removed, the facets at the same levels were removed, and the dura mater was longitudinally incised in the midline and then excised at T8	–	Basso–Beattie–Bresnahan (BBB) test and histological evaluation	Chitosan tube without growth factors or nogo–66 receptor protein	12 weeks	The combination of nogo–66 receptor protein, growth factors, and NSPCs increases the survival of transplanted NSPCs and enhances axonal regeneration	Guo et al. (2012)
Chitosan tube	Bone marrow stromal cells (BMSCs)	–	Rats	Complete spinal cord transection	–	Basso–Beattie–Bresnahan (BBB) test, retrograde tracing, and histological evaluation	Chitosan tube alone	12 weeks	Enhanced axonal regrowth, remyelination, and functional recovery	Chen et al. (2011)
Chitosan tube	Neural stem/progenitor cells (NSPCs)	Poly-lactic-co-glycolic acid (PLGA) microspheres containing dibutyryl cyclic–	Female Sprague–Dawley rats	Complete spinal cord transection laminectomy was performed on T7–T9 exposing the spinal cord. The	–	Functional and histological evaluation	Chitosan tube containing untreated NSPCs	6 weeks	dcb–AMP treatment results in greatest number of NSPCs differentiated into neurons.	Kim et al. (2011)

Continued

Table 1.3 Types of supportive cells used to enhance chitosan tubes in spinal cord repair—cont'd Combination with cells

Nerve tube	Cell type	Internal filler	Animal	Injury	Gap size (mm)	Methods	Controls	Follow-up	Outcome	Authors and year
		AMP (dbcAMP) and NSPCs		facets of the vertebrae at T7–T9 were also removed					Combination of NSPCs with chitosan tube results in extensive axonal regeneration into the injury site and improvement of functional recovery	Bozkurt et al. (2010)
Chitosan tube	Neural stem cells (NSCs)	–	Female Sprague–Dawley rats	Laminectomy at T7–T9 and spinal cord compression at T8 with a 35-g clip for 1 min	5	Basso–Beattie–Bresnahan (BBB) test and histological evaluation	NSCs transplanted without chitosan channel	6 weeks	Chitosan channels enhance the survival of transplanted NSCs. There is no difference in functional recovery between treatment and control group	Bozkurt et al. (2010)
Chitosan tube	Radial glial cells	–	Female Sprague–Dawley rats	Complete spinal cord transection at T8 level	3	Histological evaluation	–	14 weeks	After 14 weeks, radial glial cells are organized in longitudinal cord. Axons regenerate across the chitosan tube	Nomura et al. (2010)

Chitosan–alginate tube	Bone marrow stromal cells (BMSCs)	Chitosan–alginate freeze-dried sponge scaffold	Female Sprague–Dawley rats	Hemitransection at the T9 level	—	Basso–Beattie–Bresnahan (BBB) test, germ agglutinin–horseradish peroxidase retrograde tracing and histological evaluation	Direct suture of spinal dura mater after the lesion	6 weeks	Chitosan–alginate scaffolds combination with BMSCs result in better functional recovery and axonal regrowth in comparison with control group	Wang et al. (2010) and Wang, Wen, Lan, and Li (2010)
Chitosan tube	Neural stem cells (NSCs)	Multicellular sheets	Rats	Complete transection model of spinal cord injury	—	Histological evaluation	NSCs transplanted without chitosan channel	5 weeks	*In vivo* survival of NSCs and differentiation into astrocytes and oligodendrocytes. Host neurons were identified in the tissue bridge formed within the chitosan tubes	Zahir et al. (2008)

Table 1.4 Types of neurotrophic and neuroprotective factors used to enhance chitosan tubes in spinal cord repair

Combination growth factors/neuroprotective molecules

Nerve tube	Growth factor(s)/ neuroprotective molecules	Internal filler	Animal	Injury	Gap size (mm)	Methods	Controls	Follow-up	Outcome	Authors and year
Laminin-coated chitosan tube	Nogo-66 receptor protein, basic fibroblast growth factor (bFGF), epidermal growth factor (EGF), and platelet-derived growth fator (PDGF)	NSPCs preseeded on laminin-coated chitosan tube in combination with nogo-66 receptor protein, basic fibroblast growth factor (bFGF), epidermal growth factor (EGF), and platelet-derived growth fator (PDGF)	Female Sprague–Dawley rats	Complete spinal cord transection T7–T9 laminae were removed, the facets at the same levels were removed and the dura mater was longitudinally incised in the midline and then excised at T8	–	Basso–Beattie–Bresnahan (BBB) test and histological evaluation	Chitosan tube without growth factors or nogo-66 receptor protein	12 weeks	The combination of nogo-66 receptor protein, growth factors, and NSPCs increases survival of transplanted NSPCs and enhances axonal regeneration	Guo et al. (2012)
Chitosan tube	Dibutyryl cyclic-AMP	Poly-lactic-*co*-glycolic acid (PLGA) microspheres containing dibutyryl cyclic-AMP (dbcAMP) in combination with NSPCs	Female Sprague–Dawley rats	Complete spinal cord transection laminectomy was performed on T7–T9 exposing the spinal cord. The facets of the vertebrae at T7–T9 were also removed	–	Functional and histological evaluation	Chitosan tube containing untreated NSPCs	6 weeks	dcb-AMP treatment results in greatest number of NSPCs differentiated into neurons. Combination of NSPCs with chitosan tube results in extensive axonal regeneration into the injury site and improvement of functional recovery	Kim et al. (2011)

Chitosan microspheres	Atorvastatin calcium	–	Sprague–Dawley rats	Laminectomy at T7–T9 level	–	Functional evaluation using inclined plane technique of Rivlin and Tator and a modified version of the Tarlov Grading Scale and histological evaluation	Implantation of empty chitosan microspheres	5 days	Chitosan microspheres containing atorvastatin improve functional recovery, attenuate the expression of TNF-α, IL-1 beta, and IL-6, decrease lipid peroxidation levels and preserve cellular uniformity	Eroglu et al. (2010)

In another study, a laminin-coated conduit was shown to enable axons to cross the lesioned area of the spinal cord and to reduce glial scar formation (Cheng et al., 2007). Behavioral analyses evaluating the Basso–Beattie–Breshnahan motor behavior score, the sensorimotor combined behavior score, open-field walking scores, and treadmill analyses demonstrated that following the implantation of the laminin-coated nerve conduit the rats showed a tendency toward behavior improvement and functional recovery (Cheng et al., 2007). Histological and immunocytochemical analyses indicated that the implanted nerve conduit groups were capable of leading the damaged axons through the lesioned area without triggering inflammation or apoptosis (Cheng et al., 2007).

Other cell adhesion molecules, such as L1, have been shown to enhance CNS regeneration, and Xu et al., by using the optic nerve transection animal model, showed that polyglycolic acid (PGA)–chitosan conduits coated with recombinant L1–Fc have a potential role in promoting nerve regeneration by guiding axonal regrowth and remyelination (Xu et al., 2004).

3.2. Chitosan conduits combined with cells for CNS repair

NSPCs, bone marrow mesnchymal stem cells (BMSCs), and radial glial cells have been used in combination with chitosan for SCI repair (Table 1.3). Results showed that, compared to direct NSPCs injection, chitosan channels improved their survival after implantation (Bozkurt et al., 2010; Guo et al., 2012; Kim et al., 2011). In other studies, NSPCs isolated from the subependyma of lateral ventricles of adult green fluorescent protein (GFP) transgenic rat forebrains (Zahir et al., 2008) or derived from brain or spinal cord of transgenic GFP rats (Nomura, Zahir, et al., 2008) in combination with chitosan channels were implanted into the spinal cord after transection injury. These *in vivo* studies showed excellent survival of NSPCs as well as differentiation into astrocytes and oligodendrocytes (Nomura, Zahir, et al., 2008; Zahir et al., 2008). Moreover, host neurons were identified in the tissue bridge that formed within the chitosan tubes and bridged the transected cord stumps (Zahir et al., 2008). The excellent *in vivo* survival of the NSPCs coupled with their differentiation and maintenance of host neurons in the regenerated tissue bridge demonstrates that the use of three-dimensional chitosan scaffolds combined with adult spinal cord-derived NSPCs is a promising therapeutic strategy for stem cell delivery and enhances regenerative potential restoring spinal cord function after SCI although functional outcome recovery remains poor (Bozkurt et al., 2010; Nomura, Zahir, et al., 2008; Zahir et al., 2008).

BMSCs (Chen et al., 2011; Wang, Wen, Lan, & Li, 2010) and radial glial cells (Nomura et al., 2010), in combination with chitosan scaffold, have been successfully used to promote SCI repair.

3.3. Chitosan conduits combined with neurotrophic factors or neuroprotective molecules for CNS repair

Chitosan conduits combined with neurotrophic factors or neuroprotective molecules have been used for SCI repair as summarized in Table 1.4.

Recently, the neuroprotective effects of Atorvastatin, a drug used as a cholesterol lowering agent in patients, are becoming the focus of many research studies. Interestingly, chitosan microspheres containing Atorvastatin calcium have been successfully used to improve the functional outcome in an experimental SCI model (Eroglu et al., 2010). Moreover, nogo-66 receptor protein, basic fibroblast growth factor (bFGF), EGFs, and platelet-derived growth factor have been successfully used in combination with NSPCs for SCI repair (Guo et al., 2012).

4. CHITOSAN FOR PERIPHERAL NERVOUS SYSTEM REPAIR

The clinical treatment of large peripheral nerve defects requires bridging the defect that is usually accomplished by means of an autologous nerve graft. However, nerve autografting has various drawbacks such as sacrificing of a healthy functioning nerve resulting in donor site morbidity, size and quality mismatch, and possible neuroma formation at the donor site. Allografts using nerves from other individuals or animals require an additional immunosuppressant treatment. Various artificial materials have been used as scaffolds for nerve regeneration including chitosan, either alone or in combination with other materials.

In this context, chitosan is an attractive material because of its mechanical strength, porosity, biodegradability, and biocompatibility, and thus, it has been recently used for repairing nerve injury, either alone or in combination with other biomaterials (Table 1.5), adhesion molecules (Table 1.6), cells (Table 1.7), or growth factors (Table 1.8).

A number of *in vivo* studies suggested that chitosan conduits are promising candidates as supporting material for tissue engineering application in peripheral nerve reconstruction (Huang, Lu, et al., 2010; Ishikawa et al., 2007; Lauto et al., 2007, 2008; Marcol et al., 2011; Matsumoto et al., 2010; Patel et al., 2006; Rickett et al., 2011; Rosales-Cortes, Peregrina-Sandoval, Banuelos-Pineda,

Table 1.5 Chitosan conduits and luminal framework modifications used in peripheral nerve repair

Intrinsic framework

Nerve tube	Framework	Internal filler	Animal	Nerve	Injury	Gap size (mm)	Methods	Controls	Follow-up	Outcome	Authors and year
Chitosan/ polyglycolic acid tube	—	—	55-year-old male human patient	Median nerve	Nerve discontinuity	30	Compound muscle action potential recording and ninhydrin test	—	3 years	Recovery of palm abduction of the thumb and thumb–index digital opposition; reproducible compound muscle action potential of the right abductor pollicis	Gu et al. (2012)
—	Microcrystallic chitosan gel	—	Male Wistar rats	Sciatic nerve	Nerve transection	10	Autotomy behavior and histological evaluation	Nerve transection without chitosan gel implantation	20 weeks	In chitosan group neuroma formation, extraneural fibrosis are reduced in comparison to control group. There is no difference in autotomy behavior between groups	Marcol et al. (2011)
Chitosan membranes	Porous structure	—	Female Wistar rats	—	Subcutaneous implantation	—	Histological evaluation	—	8 weeks	Chitosan membranes owing their porous structure and chemical modifications and display high affinity to cellular systems	Simoes et al. (2011)

Chitosan/polyvinyl alcohol (PVA)	—	Macaques	Radial nerve	Nerve transection	20	Electrophysiological and histological evaluation	Autograft (positive control) nongrafted (negative control)	8 months	Axonal regrowth and remyelination. Recovery of compound muscle axon potential	Liu, Hou, Lin, and Wei (2011)
Chitosan tube	Tube wall with longitudinally or randomly oriented pores	—	Rats	Sciatic nerve	Nerve transection	15	Electrophysiology, retrograde labeling, and histological evaluation	Autograft	4 weeks	Axonal regeneration and motor functional recovery are improved by electrical stimulation in animals that received longitudinal pore tube compared to control group	Huang, Hu, et al. (2010) and Huang, Lu, et al. (2010)
Chitosan nano-/microfiber mesh tube	Hollow tube	—	Beagle dogs	Thoracic sympathetic nerve and phrenic nerve	Nerve transection	10	Skin temperature measurement, X-ray imaging	—	12 months	Improvement of nerve regeneration; restoration of diaphragm mobility	Matsumoto, Kaneko, Oda, and Watanabe (2010)
Freeze-dried chitosan type-III membrane and freeze-dried chitosan type-III tube	End-to-end neurorrhaphy enwrapped by chitosan membrane, 10-mm autograft enwrapped by chitosan membrane and 10-mm autograft enwrapped by chitosan tube guides	—	Sprague–Dawley rats	Sciatic nerve	End-to-end neurorrhaphy and nerve transection	10	Extensor postural thrust (EPT), withdrawal reflex latency (WRL), and ankle kinematics	10-mm nerve autograft, 10-mm PLGA tube and end-to-end surgery alone	20 weeks	Better nerve regeneration in chitosan type III tubulization group than PLGA tubulization control group	Simoes et al. (2010)

Continued

Table 1.5 Chitosan conduits and luminal framework modifications used in peripheral nerve repair—cont'd
Intrinsic framework

Nerve tube	Framework	Internal filler	Animal	Nerve	Injury	Gap size (mm)	Methods	Controls	Follow-up	Outcome	Authors and year
Microporous–chitosan tube	Filaments	Polyglycolicacid (PGA) oriented filaments	Female Sprague–Dawley rats	Sciatic nerve	Nerve transection	10	Electrophysiological and histological evaluation	10-mm autograft (positive control) and nongrafted (negative control)	6 months	Better axonal regeneration, SC myelination, and reinnervation of atrophic denervated muscle in the chitosan/PGA graft group	Jiao et al. (2009)
Bilayered chitosan nonwoven nanofibers mesh tube	Filaments	Inner layer of orientated nanofibers and outer layer of random nanofibers	Male Sprague–Dawley rats	Sciatic nerve	Nerve transection	10	Von Frey hair test, Static toe spread factor (STSF), electrophysiological and histological evaluation	Autograft (positive control) and random nanofibers chitosan mesh tube (negative control)	30 weeks	Functional recovery in the bilayered chitosan nonwoven nanofibers mesh tube group matches autograft group. Sprouting of myelinated axons and axonal maturation occurs in both groups at the same level	Wang et al. (2009)
Poly-D-L-lactic acid/chondroitin sulfate/chitosan tube (PDLLA/CS/CHS)	–	–	Rats	–	–	–	–	PDLLA tube	3 months	The PDLLA/CS/CHS tube allows nerve regeneration without acidity-caused irritation and acidity-induced autoaccelerating degradation behavior typical of PDLLA alone	Xu et al. (2009)

Microporous–chitosan tube	Filaments	Polyglycolic acid (PGA) oriented filaments	37–years-old male human patient	Median nerve	Nerve discontinuity	35	JAMAR Hand evaluation Kit (5030KIT), touch test sensory evaluators, disk-criminator, electrophysiological evaluation, blood and urinary test, and serum biochemical examination	–	3 years	Motor and sensory function recovery	Fan et al. (2008)
Laser-activated chitosan adhesive	–	–	Rats	Tibial and sciatic nerve	Sutureless anastomosis	–	Histological evaluation	–	3 days	Successful nerve anastomosis; myelinated axons display normal number and morphology	Lauto et al. (2008)
Bilayered chitosan film tube	Nano–/ microfibers	Nano–/ microfibers mesh tube with a DAc of 78% or 93% and film tube with a DAc of 93%	Male Sprague–Dawley rats	Sciatic nerve	Nerve transection	15	Von Frey hair test, Static toe spread factor (STSF), electrophysiological and histological evaluation	15-mm autograft	10 weeks	Functional recovery of motor activity delays in each group compared to autograft group; good sensory and functional recovery in autograft group followed by nano–/ microfibers mesh tubes with a DAc of 93% group	Wang et al. (2008) and Wang, Itoh, Matsuda, Ichinose, et al. (2008)

Continued

Table 1.5 Chitosan conduits and luminal framework modifications used in peripheral nerve repair—cont'd Intrinsic framework

Nerve tube	Framework	Internal filler	Animal	Nerve	Injury	Gap size (mm)	Methods	Controls	Follow-up	Outcome	Authors and year
Polylactic acid (PLA)/ chitosan tube	Hollow tube	PLGA filler	Male Sprague–Dawley rats	Sciatic nerve	Nerve transection	10	Electrophysiological and histological evaluation	10-mm autograft and 10-mm silicone conduit group	12 weeks	Axonal regeneration, sciatic nerve functional recovery, and muscle reinnervation of chitosan–PLA group are close to control group	Xie, Li, Gu, Liu, and Shen (2008)
Chitosan	Microgrooved polymer	–	Male Sprague–Dawley rats	Sciatic nerve	Nerve transection	12	Histological evaluation	12-mm silicone group and smooth chitosan conduits	6 weeks	Microgrooved conduits enhance peripheral nerve regeneration in comparison with the smooth conduits	Hsu et al. (2007)
Chitosan tube	Sponge	Freeze-dried chitosan gel sponge	Male Wistar rats	Sciatic nerve	Nerve transection	8	Histological evaluation	8-mm gap without chitosan tube implantation	4 months	Axonal regeneration and remyelination	Ishikawa et al. (2007)
Chitosan tube	Hollow tube	–	Rats	Sciatic nerve	Nerve transection	10	Walking track analysis and histological evaluation	10-mm autograft	12 weeks	Decrease in muscle atrophy, increase in axonal growth and functional recovery	Patel et al. (2006)

Microporous chitosan tube	Filaments	Polyglycolic acid (PGA)–oriented filaments	Beagle dogs	Sciatic nerve	Nerve transection	30	Electrophysiological, histological evaluation and FluoroGold retrograde tracing	30-mm autograft group (positive control) and nongrafted group (negative control)	6 months	Restoration of nerve continuity, functional recovery, and target muscle reinnervation in the chitosan PGA graft group	Wang et al. (2005)
Chitosan/PLA tube	—	—	Rats	Sciatic nerve	Nerve transection	5	Electrophysiological and histological evaluation	Silicon conduits and autograft group	12 weeks	Chitosan/PLA tube results in nerve regeneration, axon's quality, and quantity close to autograft's group	Xie, Li, and Zhao (2005)
Chitosan tube	Hollow tube	—	Female Beagle dogs	Sciatic nerve	Axotomy	—	IgG and IgM serum analysis	Intact and axotomized control groups	60 days	Chitosan implants do not affect the immune response	Rosales-Cortes, Peregrina-Sandoval, Banuelos-Pineda, Castellanos-Martinez, et al. (2003) and Rosales-Cortes, Peregrina-Sandoval, Banuelos-Pineda, Sarabia-Estrada, et al. (2003)
Apatite/chitosan tube	Hollow tube	—	Male Sprague–Dawley rats	Sciatic nerve	Nerve transection	10	Histological evaluation	10-mm chitosan tube	12 weeks	Apatite-treated chitosan hollow tube keeps its shape in vivo and induces nerve regeneration	Yamaguchi et al. (2003)

Table 1.6 Modifications to chitosan conduits surface used in peripheral nerve repair

Surface modification

Nerve tube	Surface modification	Internal filler	Animal	Nerve	Injury	Gap size (mm)	Methods	Controls	Follow-up	Outcome	Authors and year
Chitosan nanofibers mesh tube	Coating with polarized and nonpolarize-β-tricalcium phosphate particles	–	Male Wistar rats	Sciatic nerve	Nerve transection	10	Static toe spread factor (STFS), von Fray hair test, electrophysiological and histological evaluation	Autograft	12 weeks	Motor and sensory nerve function and electrophysiological recovery progress with time in each group. Immunofluorescence reveals more rapid nerve regeneration in the polarized tube group compared with the nonpolarized tube group. The axon density and axon area in the polarized tube group are significantly greater than those in the chitosan mesh tube and nonpolarized group, and it shows no significant differences from the control group	Wang, Itoh, et al. (2010) and Wang, Wen, Lan, and Li (2010)
Chitosan–collagen tube	Blending with collagen	Longitudinally oriented microchannels	Rats	Sciatic nerve	Nerve transection	15	Functional and histological evaluation	Autograft	–	Chitosan–collagen scaffolds achieve nerve regeneration and functional recovery equivalent to autograft.	Hu et al. (2009)

Chitosan–collagen tube	Blending with collagen	–	Rats	Sciatic nerve	Nerve transection	–	Gait analysis and behavioral test	Unblended chitosan nerve guide	12 weeks	Collagen-blended chitosan nerve guide enhances motor and sensory recovery compared with unblended nerve guides	Patel et al. (2008a, 2008b)
Chitosan film	Conjugation with C(G)YIGSR peptide from laminin-1	Chitosan nonwoven nano-/microfibers mesh	Male Sprague–Dawley rats	Sciatic nerve	Nerve transection	10	Static toe spread factor (STSF) and histological evaluation	Autograft	10 weeks	Nerve regeneration in the chitosan/CYIGSR group is similar to the autograft	Wang, Itoh, Matsuda, Aizawa, et al. (2008) and Wang, Itoh, Matsuda, Ichinose, et al. (2008)
Thiolated and nonthiolated hydroxyapatite-coated chitosan tube	Adsorbed YIGSR peptide from laminin-1	–	Male Sprague–Dawley rats	Sciatic nerve	Nerve transection	15	Electrophysiological and histological evaluation	Autograft	12 weeks	YIGSR peptide enhances nerve regeneration and axonal sprouting from the proximal stump to the distal one	Itoh et al. (2005)
Circular and triangular cross-section chitosan tube	Absorption of laminin and YIGSR and IKVAV laminin–peptide	–	Male Sprague–Dawley rats	Sciatic nerve	Nerve transection	15	Electrophysiological and histological evaluation	Autograft	12 weeks	Triangular tubes have higher mechanical strength and inner volume than circular ones. SC migration and axonal outgrowth is enhanced. Nerve tissue regeneration occurs in both laminin and laminin–peptide groups	Itoh, Suzuki, et al. (2003)

Continued

Table 1.6 Modifications to chitosan conduits surface used in peripheral nerve repair—cont'd

Surface modification

Nerve tube	Surface modification	Internal filler	Animal	Nerve	Injury	Gap size (mm)	Methods	Controls	Follow-up	Outcome	Authors and year
Circular and triangular cross-section chitosan tube coated with hydroxyapatite	Absorption of laminin and YIGSR and IKVAV laminin–peptide	—	Male Sprague–Dawley rats	Sciatic nerve	Nerve transection	15	Electrophysiological and histological evaluation	Autograft	8 weeks	Triangular tubes coated with hydroxyapatite have higher mechanical strength and inner volume than circular ones. Nerve tissue regeneration occurs in both laminin and laminin–peptide groups matching isograft group. Functional recovery is delayed	Itoh, Yamaguchi, et al. (2003)
Triangular cross-section chitosan tube	Covalent binding of laminin and YIGSR and IKVAV laminin–peptide	—	Male Sprague–Dawley rats	Sciatic nerve	Nerve transection	15	Electrophysiological and histological evaluation	Autograft	8 weeks	YIGSR, followed by IKVAV, laminin–peptide matches the effectiveness of intact laminin in enhancing nerve regeneration	Suzuki et al. (2003)
Chitosan–collagen film	Blending with collagen	—	Rats	Sciatic nerve	Nerve transection	5 or 10	Electrophysiological and histological evaluation	Autograft	12 weeks	In 5-mm defects, nerve regeneration is similar to control group. In 10-mm defects, nerve regeneration is inferior to control group. Chitosan–collagen film conduits are degraded at 12 weeks postsurgery	Wei et al. (2003)

Table 1.7 Types of supportive cells used to enhance chitosan tubes in peripheral nerve repair Combination with cells

Nerve tube	Cell type	Internal filler	Animal	Nerve	Injury	Gap size (mm)	Methods	Controls	Follow-up	Outcome	Authors and year
Silicon tube	Autologous bone marrow mesenchymal stem cells (BMSCs)	Laminin-modified chitosan film with BMSCs	Female Sprague–Dawley rats	Sciatic nerve	Nerve transection	10	Histological evaluation and FluoroGold retrograde tracing	Empty silicon tube	16 weeks	BMSCs containing scaffolds improve nerve regrowth, muscle mass maintenance, and functional recovery.	Hsu et al. (2013)
Chitosan tube	Autologous bone marrow mesenchymal stem cells (BMSCs)	Longitudinally aligned poly lactic-co-glycolic acid (PLGA) fibers	Rhesus monkeys	Median nerve	Nerve transection	50	Electrophysiological and histological evaluation, FluoroGold retrograde tracing	Chitosan–PLGA fibers alone	12 months	Twelve months after grafting, nerve function recovery, and morphological reconstruction of BMSCs containing PLGA/chitosan scaffold is superior to that of chitosan/PLGA scaffold alone	Hu et al. (2013)
Chitosan tube	Autologous bone marrow mesenchymal stem cells (BMSCs)	Poly lactic-co-glycolic-acid (PLGA) oriented fibers	Male Beagle dogs	Sciatic nerve	Nerve transection	60	Electrophysiological, retrograde fluorogold tracing, and histological evaluation	Autograft and chitosan/PLGA fibers alone	12 months	MSCs in combination with chitosan/PLGA fibers tube improve nerve repair in comparison to chitosan/PLGA fibers tube alone	Xue et al. (2012)
Chitosan tube	Bone marrow stromal cells (BMSCs)–derived Schwann cells or sciatic nerve–derived Schwann cells (SCs)	—	Rats	Sciatic nerve	Nerve transection	8	Sciatic nerve function index (SFI) and histological evaluation	Chitosan tube alone	6 weeks	Six weeks postsurgery, the SFI, average regenerated fiber density, and fiber diameter in nerve bridged with BMSCs are similar to autograft	Zheng and Cui (2012)

Continued

Table 1.7 Types of supportive cells used to enhance chitosan tubes in peripheral nerve repair—cont'd Combination with cells

Nerve tube	Cell type	Internal filler	Animal	Nerve	Injury	Gap size (mm)	Methods	Controls	Follow-up	Outcome	Authors and year
Chitosan tube	Bone marrow stromal cells (BMSCs)–derived Schwann cells or sciatic nerve–derived Schwann cells (SCs)	Matrigel	Male Sprague–Dawley rats	Sciatic nerve	Nerve transection	12	Footprint analysis, compound muscle action potential (CMAP) measurements, histological evaluation	Autograft and PBS-filled conduits	3 months	Nerve conduction velocity, average regenerated myelin area, and myelinated axon count in nerve bridged with BMSCs-derived SCs are similar to those treated with sciatic nerve–derived SCs and higher than those bridged with PBS-filled conduits	Ao et al. (2011)
Chitosan/polyglycolic acid tube	Autologous bone marrow mesenchymal stem cells (BMSCs)	–	Beagle dogs	Sciatic nerve	Nerve transection	50	Electrophysiological and histological evaluation, FluoroGold retrograde tracing	Autograft	6 months	Introduction of BMSCs in the conduits promotes nerve regeneration and functional recovery	Ding et al. (2010)
Chitosan-3-glycidoxypropyltrimethoxysilane (GPTMS) cross-linked membranes	Predifferentiated N1E-115 cells	Chitosan membrane covered with a cell monolayer	Female Wistar rats	Sciatic nerve	Crush injury	10	Kinematic analysis and histological evaluation	Unoperated animals	8 weeks	Local enwrapping with chitosan membrane without N1E-115 cells improves axonal regrowth and functional recovery	Amado et al. (2008)
Chitosan–Au-nanocomposites	Neural stem cell (NSC)	–	Male Sprague–Dawley rats	Sciatic nerve	Nerve transection	10	Histological evaluation	Chitosan–Au-nanocomposites tube without NSC	6 weeks	In comparison to control group, the number of regenerated axons, the regenerated area, and the number of blood vessels are significantly higher in the NSCs preseeded tube group	Lin, Jen, Hsu, and Chiu (2008)

Table 1.8 Types of neurotrophic and neuroprotective factors used to enhance chitosan tubes for peripheral nerve repair

Neurotrophic factors

Nerve tube	Growth factor(s)	Carrier/delivery system	Animal	Nerve	Injury	Gap size (mm)	Methods	Controls	Follow-up	Outcome	Authors and year
Chitosan tube	Nerve growth factor (NGF)	NGF immobilization via genipin cross-linking	Rats	Sciatic nerve	Nerve transection	10	Electrophysiological and histological evaluation	Autograft	24 weeks	Chitosan/NGF tube promotes nerve regeneration close to the autograft group	Wang et al. (2012)
Poly-D, L-lactic–acid/chondroitin sulfate/chitosan tube	NGF	Immobilization on the tube surface	Sprague–Dawley rats	Sciatic nerve	Nerve transection	10	Electrophysiological and histological evaluation	Autograft	6 months	No connective tissue in growth. Rapid functional recovery and nontoxicity of degradation products. NGF promotes nerve regeneration close to the autograft group	Xu, Yan, and Li (2011)
Chitosan tube	Immunophilin ligand FK506	Drug loading into the semipermeable wall of chitosan tube	Rats	Sciatic nerve	Nerve transection	–	Electrophysiological and histological evaluation	Autograft	8 weeks	FK506 treatment results in more mature appearance of myelinated fibers. The amplitude and velocity of compound muscle action potential of treated group are close to the autograft group	Li et al. (2010)

Continued

Table 1.8 Types of neurotrophic and neuroprotective factors used to enhance chitosan tubes for peripheral nerve repair—cont'd Neurotrophic factors

Nerve tube	Growth factor(s)	Carrier/delivery system	Animal	Nerve	Injury	Gap size (mm)	Methods	Controls	Follow-up	Outcome	Authors and year
Chitosan tube	Basis fibroblast growth factor (bFGF)	Heparin-incorporated fibrin–fibronectin matrix	Rats	Sciatic nerve	Nerve transection	10	Conduction velocity recovery index (CVRI), muscle restoration rate (MRR), and histological evaluation	Autograft	3 months	CVRI and MRR in animals of bFGF group are similar to those of autograft group	Han, Ao, Chen, Wang, and Zuo (2010)
Poly lactic-co-glycolic acid (PLGA)/chitosan tube	CNTF	CNTF blending with chitosan	Cross-bred dogs	Tibial nerve	Nerve transection	25	Electrophysiological and histological evaluation	PLGA/chitosan tube without CNTF	3 months	PLGA/chitosan–CNTF tube promotes nerve regeneration close to the autograft group	Shen et al. (2010)
—	Chitooligosaccharides (COSs)	Intravenous injections of 1.5 or 3 mg/kg body weight of COSs over 6-week period	New Zealand rabbits	Common peroneal nerve	Nerve crush injury	20	Electrophysiological and histological evaluation	Saline injections	6 weeks	Compound muscle action potentials, number of regenerated myelinated nerve fibers, thickness of regenerated myelin sheaths and the cross-sectional area of	Gong, Gong, Gu, and Ding (2009)

Material	Agent	Method	Animal	Nerve	Injury	Number	Evaluation	Control	Time	Results	Reference
										tibialis posterior muscle fibers are significantly improved in animals treated with COSs in a dose-dependent manner. COSs display a neuroprotective effect	Jiang, Zhuge, Yang, Gu, and Ding (2009)
—	Chitooligosaccharides (COSs)	Intraperitoneally injections of 3 or 6 mg/kg body weight of COSs over 3-week period	Sprague–Dawley rats	Sciatic nerve	Nerve crush injury	—	Electrophysiological, measurement of withdrawal reflex latency (WRL), walking track analysis, and histological evaluation	Saline injections	3 weeks	COSs promote peripheral nerve regeneration and functional recovery of injured nerves. COSs display a neuroprotective effect	
Chitosan tube	GDNF	GDNF blending with chitosan	Lewis rats	Sciatic nerve	Nerve transection	10	Histological evaluation	Autograft	12 weeks	Chitosan–GDNF tube enhances nerve regeneration process during the initial stages of nerve repair	Patel, Mao, Wu, and VandeVord (2009)
Chitosan tube	GDNF	GDNF blending with chitosan	Lewis rats	Sciatic nerve	Nerve transection	10	Functional gait analysis and sensitivity test	Autograft	12 weeks	Chitosan–GDNF tube increases functional recovery compared to unblended chitosan groups	Patel, Mao, Wu, and Vandevord (2007)

Continued

Table 1.8 Types of neurotrophic and neuroprotective factors used to enhance chitosan tubes for peripheral nerve repair—cont'd Neurotrophic factors

Nerve tube	Growth factor(s)	Carrier/delivery system	Animal	Nerve	Injury	Gap size (mm)	Methods	Controls	Follow-up	Outcome	Authors and year
Chitosan tube	Progesterone (PROG) and pregnenolone (PREG)	GDNF blending with chitosan	New Zealand rabbits	Facial nerve	Nerve transection	10	Histological evaluation	Chitosan tube without PROG/PREG	45 days	Chitosan–PROG/PREG tube promotes axonal regeneration and myelination	Chavez-Delgado et al. (2005)
Chitin tube	Nerve growth factor (NGF)	–	Sprague–Dawley rats	Sciatic nerve	Nerve transection	5	Electrophysiological and histological evaluation	Epineurium direct suture	8 weeks	The repair effects of chitin conduit bridging peripheral nerve with 5-mm gap are better than epineurium suture directly, and possess the potential to substitute the epineurium suture	Zhang et al. (2005)
Chitosan tube	Progesterone (PROG) and pregnenolone (PREG)	GDNF blending with chitosan	New Zealand rabbits	Facial nerve	Nerve transection	10	Histological evaluation	Chitosan tube without PROG/PREG	45 days	Chitosan–PROG/PREG tube promotes myelination	Chavez-Delgado et al. (2003)

Chitosan tube	Progesterone (PROG)	–	Female dogs	Sciatic nerve	Nerve transection	15	Histological evaluation	Chitosan tube without PROG	–	Chitosan/PROG tube promotes axonal regeneration and myelination	Rosales-Cortes, Peregrina-Sandoval, Banuelos-Pineda, Castellanos-Martinez, et al. (2003) and Rosales-Cortes, Peregrina-Sandoval, Banuelos-Pineda, Sarabia-Estrada, et al. (2003)

Sarabia-Estrada, et al., 2003; Simoes et al., 2011; Wang, Itoh, Matsuda, Ichinose, et al., 2008; Wang, Itoh, et al., 2010; Wang et al., 2009; Yamaguchi et al., 2003; Zhang et al., 2005).

An experimental study reconstructing 10-mm gaps in the rat sciatic nerve showed that chitosan tubes induce nerve regeneration and are gradually degraded and absorbed *in vivo* (Yamaguchi et al., 2003). Patel et al. (2006) reported that chitosan nerve guides improve functional nerve recovery, by increasing axonal growth, reduce muscle atrophy, and restore functional strength.

It has also been shown that the regeneration of the axotomized dog sciatic nerve can be improved through tubulization with chitosan without affecting the immune response (Rosales-Cortes, Peregrina-Sandoval, Banuelos-Pineda, Sarabia-Estrada, et al., 2003).

As an internal conduit framework is concerned, a freeze-dried chitosan gel sponge has been used to bridge a 8-mm gap lesion in the rat sciatic nerve; 14 days after the surgery, the regenerated nerve fibers are extended inside the conduit along a cell layer provided by infiltrating cells, and 2 months post-surgery, the regenerated nerve appeared well remyelinated, indicating that the chitosan gel sponge material might be a promising graft for peripheral nerve regeneration (Ishikawa et al., 2007).

Chitosan nano-/microfiber mesh tubes have been safely used also to successfully regenerate damaged thoracic nerves in beagle dogs, specifically sympathetic and phrenic nerve, resulting in restoration of the respiratory function (Matsumoto et al., 2010).

Chitosan nanofibers mesh tubes with or without orientation and bilayered chitosan mesh tubes with an inner layer of oriented nanofibers and an outer layer of randomized nanofibers have been used to bridge sciatic nerve defects in rats. Sprouting of axons and axonal maturation followed by functional recovery occurred in the oriented conduits as well as in the bilayered conduits matching the outcome of the nerve autografts (Wang et al., 2009).

Nano-/microfiber mesh tubes with a DD of 78% or 93% investigated in the 10-mm rat sciatic nerve gap repair resulted in better sensory recovery for mesh tubes with a DD of 93%. These tubes have adequate mechanical properties to preserve the tubes internal lumen, resulting in better cell migration and adhesion as well as humoral permeation enhancing nerve regeneration (Wang, Itoh, Matsuda, Ichinose, et al., 2008).

Interestingly, a chitosan-based-laser-activated adhesive has been successfully applied to perform sutureless coaptation of the rat tibial nerve without altering axon number and morphology (Lauto et al., 2007, 2008). A photocrosslinkable hydrogel based on chitosan has been *in vitro* successfully characterized and proposed as a new adhesive for peripheral nerve anastomosis (Rickett et al., 2011).

Chitosan has further been used in combination with other biomaterials for bridging peripheral nerve gaps (Fan et al., 2008; Gu et al., 2012; Jiao et al., 2009; Lin et al., 2008; Liu et al., 2011; Simoes et al., 2010; Wang et al., 2005; Xie et al., 2005; Xie et al., 2008; Xu et al., 2009). Chitosan–polylactid acid (PLA) composite nerve conduits showed good biocompatibility and permeability, good mechanical strength, intensity, and elasticity, facilitating microsuture manipulation, and they provide enough mechanical strength to support nerve regeneration. Chitosan–PLA conduits promoted axonal regeneration of rat sciatic nerves across a defect of 10 mm, with comparable success to nerve autografts, resulting in muscle reinnervation 12 weeks postsurgery (Xie et al., 2008).

Yet, tubular grafts made out of chitosan membranes have been successfully used to improve peripheral nerve functional recovery after neurotmesis of the rat sciatic nerve, and they induced better nerve regeneration and functional recovery when compared with poly-lactic-polyglycolic acid (PLGA) control tubes (Simoes et al., 2010).

A dual component artificial nerve conduit consisting of an outer chitosan microporous conduit and an internal oriented PGA filament matrix has been used to bridge a 10-mm defect in rats after long-term delay (3 or 6 months), resulting in reinnervation of the atrophic denervated muscle by regenerating neurites through new muscle–nerve connections (Jiao et al., 2009). The same conduit has been used to regenerate 30-mm beagle dog sciatic nerve defects, resulting in reconstruction and restoration of nerve continuity and functional recovery as indicated by improved locomotion activities of the operated limb after target muscle reinnervation (Wang et al., 2005).

Finally, a couple of clinical studies in which chitosan scaffolds have been used to repair peripheral nerve have been reported. Chitosan/PGA artificial conduits have been successfully used in repairing a 35-mm-long median nerve defect of a human patient. During the 3-year follow-up period, an ongoing motor and sensory functional recovery postimplantation was detected (Fan et al., 2008). Chitosan/PGA conduits have further been used to repair a 30-mm long median nerve defect in the right distal forearm of a 55-year-old male patient. Thirty-six months after the surgery, reproducible compound muscle action potentials have been recorded on the right abductor policies, the palm adduction of the thumb, and the thumb-index digital opposition recovered and facilitated the accomplishment of fine activities (Gu et al., 2012). These results are very promising and suggest that the chitosan/PGA artificial nerve graft could be used for clinical reconstruction of major peripheral nerves' defects in the forearm.

4.1. Surface modification of chitosan conduits for PNS repair

Various studies have been dedicated to develop scaffolds with an inner structure mimicking the nerve-guiding basal lamina present in nerve autografts. Designing biomaterial surfaces in order to mediate cellular interactions through coupling of specific cell membrane receptors may allow to control cell adhesion, cell migration, and tissue organization and consequently improve SC migration and organization and axonal outgrowth.

To enhance nerve regeneration, a number of studies, in which laminin has been used to coat the inner tube surface, have been reported (Madison, da Silva, Dikkes, Chiu, & Sidman, 1985; Yoshii, Yamamuro, Ito, & Hayashi, 1987). Although these studies proved the effectiveness of scaffold enrichment with laminin, this molecule is not easy to synthesize and it cannot be applied in human patients because it is a tumor-inducing material (Timpl et al., 1979).

To avoid these limitations, laminin peptides (YIGSR and IKVAV) have been used for functionalizing the inner surface of chitosan-based conduits too, with the goal of improving nerve regeneration. YIGSR-treated chitosan–hydroxyapatite (HAp) tubes enhanced SC migration and long-distance growth of regenerated axons in a 15-mm gap rat sciatic nerve injury model in comparison with full laminin-1-coated chitosan tubes. Histological regeneration, as well as mechanical properties, of the YIGSR-treated chitosan–HAp tubes matched with those of nerve autografts, although functional recovery was delayed (Itoh, Yamaguchi, et al., 2003).

Yet, circular and triangular cross-section chitosan tubes combined with YIGSR and IKVAV have been used to bridge a 15-mm gap in the sciatic nerve of rats. The mechanical strength of triangular tubes was higher than that of circular tubes and their inner volume tended to be larger. Again, YIGSR and IKVAV matched the effectiveness of full laminin to enhance nerve regeneration (Itoh, Suzuki, et al., 2003; Suzuki et al., 2003).

In another study, thiolated and nonthiolated HAp-coated crab tendon triangular cross-section chitosan tubes, both alone and conjugated with the

YIGSR peptide, have been utilized to bridge a 15-mm rat sciatic nerve gap. Histological and functional recovery analyses showed that while thiolation might have delayed nerve tissue regeneration, YIGSR peptides enhanced nerve regeneration by promoting sprouting from the proximal nerve stump and long-distance growth of regenerated axons throughout the tube (Itoh et al., 2005). Moreover, a bilayered chitosan tube that comprised of an outer layer of a chitosan film and an inner layer of a nonwoven chitosan nano-/microfiber mesh coated with the YIGSR peptide (in which a glycin spacer has been introduced) has been successfully used to bridge the 15-mm rat sciatic nerve gap (Wang, Itoh, Matsuda, Aizawa, et al., 2008).

Chitosan has also been used in combination with collagen for nerve repair. Collagen-blended chitosan nerve guides have been successfully tested *in vivo* in rats and enhanced both motor and sensory recoveries in comparison with unblended nerve guides (Patel et al., 2008a; Wei, Lao, & Gu, 2003). Moreover, collagen–chitosan nerve guides promote and support axonal sprouting, increase axon diameters and the area occupied by regenerated axons, and further improve axonal maturation (Patel et al., 2008b). Collagen–chitosan conduits, characterized by longitudinally orientated microchannels, have also been shown to allow good nerve regeneration and functional recovery across 15-mm-long rat sciatic nerve defects (Hu et al., 2009). A summary of the different studies of chitosan combined with adhesion molecules is reported in Table 1.6.

4.2. Chitosan conduits combined with cells for PNS repair

Directing SC migration by biomaterial substrates is receiving much attention in peripheral nerve tissue engineering (Heath & Rutkowski, 1998). SCs secrete neurotrophic factors and express cell adhesion molecules that enhance peripheral nerve regeneration (Gravvanis et al., 2004; Heath & Rutkowski, 1998; Ide, 1996). They form an endoneurial sheath which acts as a guide for axonal growth from the proximal to the distal nerve stumps. They also play a role in clearing debris and creating an appropriate *milieu* for nerve regrowth. Due to their role in nerve regeneration, the behavior of SCs on a biomaterial used for nerve conduit fabrication is clearly a key issue (Guenard, Kleitman, Morrissey, Bunge, & Aebischer, 1992; Heath & Rutkowski, 1998) and it has been shown that SCs can be aligned by culturing them on biomaterial surface groves which provide micrometric dimensions (Hsu et al., 2007; Wang et al., 2009).

Another strategy to support SC colonization inside nerve guides is pre-enrichment of the conduit with SCs or their precursors. BMSC-derived SCs have been used in combination with chitosan conduits to repair 12-mm rat sciatic nerve gap, resulting in nerve conduction velocities, average regenerated myelin area, and number of myelinated axons similar to those conduits treated with sciatic nerve-derived SCs (Ao et al., 2011).

Chitosan-based conduits combined with autologous BMSCs have been successfully utilized to bridge 8-mm-long sciatic nerve defects in adult rats (Zheng & Cui, 2012).

Chitosan/PLGA-based neuronal scaffolds, in which autologous BMSCs have been incorporated, promoted dog sciatic nerve regeneration and functional recovery across 50- to 60-mm-long gaps. The outcome was close to that of nerve autografts and better than that of chitosan/PLGA-based scaffolds alone (Ding et al., 2010; Xue et al., 2012).

Very recently, chitosan/PLGA nerve grafts combined with autologous BMSCs have been utilized to bridge 50-mm-long medial nerve defects in rhesus monkeys. Functional recovery was more efficient when chitosan/PLGA nerve grafts were combined with BMSCs instead of the used cell-free chitosan/PLGA grafts. Moreover, blood tests and histopathological examinations demonstrated that BMSCs could be safely used in primates (Hu et al., 2013).

Recently, combination of peptide and cells to enhance nerve regeneration has been used to bridge 10-mm gap lesion in rat sciatic nerve. Laminin-coated chitosan conduit in combination with BMSCs results in enhancement of nerve regrowth, muscle mass maintaining, and functional recovery. BMSCs inhibited neuronal cell death and overturn the inflammatory response induced by long-term chitosan implantation promoting nerve regeneration (Hsu et al., 2013).

Chitosan 3-glycidoxypropyl-methyldiethoxysilane-cross-linked membranes have been used for peripheral nerve reconstruction in the rat sciatic nerve model, either alone or in combination with N1E-115 NSCs. Chitosan membranes showed good biocompatibility and were suitable for N1E-115 cells growth. However, the study of *in vivo* nerve regeneration after nerve crush injury showed that freeze-dried chitosan membrane without cell enrichment improved axonal regrowth and functional recovery, suggesting that nerve enwrapping with chitosan membrane alone may be an effective method for improving peripheral nerve repair, while enrichment with N1E-115 neural cells is not (Amado et al., 2008; Simoes et al., 2011).

4.3. Chitosan conduits combined with neurotrophic factors or neuroprotective molecules for PNS repair

After an injury, one of the causes contributing to apoptosis and poor functional recovery is the neurotrophic factor deprivation (Ramer, Priestley, & McMahon, 2000). Nerve regeneration has been found to be enhanced by utilizing guidance channels filled with neurotrophic factors, such as GDNF, ciliary nerotrophic factor (CNTF), FGF-2, and NT-3 (Grothe, Haastert, & Jungnickel, 2006; Madduri, Feldman, Tervoort, Papaloizos, & Gander, 2010; Madduri, Papaloizos, & Gander, 2010; Oh et al., 2008; Pfister et al., 2008; Yang et al., 2007). Various types of nerve guides have been developed by blending chitosan with different growth factors (Table 1.8). Adding growth factors can support nerve regeneration by improving the biological properties of a nerve guide. GDNF has been mixed to chitosan–laminin conduits which have been used to bridge 10-mm rat sciatic nerve gaps. GDNF–laminin-blended chitosan conduits increased functional recovery and decreased muscle atrophy compared with unblended chitosan conduits (Patel et al., 2007). Histologically, the GDNF–laminin-blended chitosan conduits demonstrated at 6 weeks postsurgery regenerated axons with higher axonal area and myelination in comparison with control conditions. At 9–12 weeks postsurgery, however, control groups matched the GDNF–laminin-blended chitosan group indicating that these kinds of conduits exert their positive effects during the initial stages of nerve regeneration only (Patel et al., 2009).

In another study, CNTF-coated PLGA chitosan nerve conduit has been utilized to repair 25-mm-long segments in the canine tibial nerve. Histological results demonstrated that the PLGA/chitosan–CNTF conduits were capable of guiding the damaged axons through the lesioned area, resulting in good functional recovery close to the outcome in the nerve autograft group (Shen et al., 2010).

Yet, also NGF has been immobilized onto biodegradable PDLLA/CS/CHS nontoxic nerve conduits, resulting in good functional recovery after bridging 10-mm defects in the rat sciatic nerve (Xu et al., 2011). Moreover, chitosan conduit on which NGF was immobilized via genipin cross-linking resulted in nerve reconstruction and muscle reinnervation in a 10-mm-long sciatic nerve gap in rat (Wang et al., 2012).

Chitosan nerve conduits filled with heparin-incorporated fibrin–fibronectin matrix for bFGF delivery have been successfully used to repair sciatic nerve defects of 10-mm in adult rats (Han et al., 2010).

Interestingly, a couple of studies showed that chitooligosaccharides (COSs), the biodegradation products of chitosan, promote peripheral nerve regeneration and functional recovery in the rat sciatic nerve crush injury model and rabbit common peroneal nerve crush injury model, suggesting their potential application in peripheral nerve repair as neuroprotective agents (Gong et al., 2009; Jiang et al., 2009).

Moreover, immunophilin FK506 combined with biodegradable chitosan guide provides neurotrophic and neuroprotective actions, promoting nerve regeneration in a rat sciatic nerve defect model (Li et al., 2010).

Also neurosteroids, such as progesterone (PROG) and pregnenolone (PREG), have been used for potentiating conduit nerve repair as these hormones are synthesized by SCs (Baulieu & Schumacher, 2000) and induce myelination binding on intracellular receptors which activate the synthesis of myelin protein P0 and PMP22 (Desarnaud et al., 1998; Jung-Testas, Schumacher, Robel, & Baulieu, 1996). Furthermore, neurite outgrowth may be stimulated by a PROG metabolite (5-α-tetrahydroprogesterone) through GABA(A) receptors (Guennoun et al., 2001; Koenig, Gong, & Pelissier, 2000). Chitosan conduits have been used to deliver PROG into a 10-mm rabbit facial nerve gap model. The released PROG promoted nerve regeneration to a high degree so that at 45 days postsurgery myelinated fibers were observed both in the proximal and distal nerve stumps (Chavez-Delgado et al., 2003). Similar results have been reported using axotomized dog sciatic nerve injured model (Rosales-Cortes, Peregrina-Sandoval, Banuelos-Pineda, Castellanos-Martinez, et al., 2003). Chitosan tubes delivering a combination of PROG and PREG have also been tested. Fifteen days postsurgery, the regenerating tissue contained SCs holding nonmyelinated fibers, whereas at 45 days postsurgery, the regenerating tissue displayed myelinated fibers of different shape, size, and myelin sheath thickness (Chavez-Delgado et al., 2005). These findings indicate that chitosan conduits allows regeneration of nerve fibers and that long-time release of neurosteroid from the conduits induces faster regeneration.

5. CONCLUSION

Among the many different types of biomaterials that have been proposed as optimal candidates for neural repair scaffolds, chitosan has been receiving growing interests among basic and clinical scientists. In this review, we have overviewed the more relevant articles demonstrating that this biomimetic material exerts a positive effect on cell cultures of both

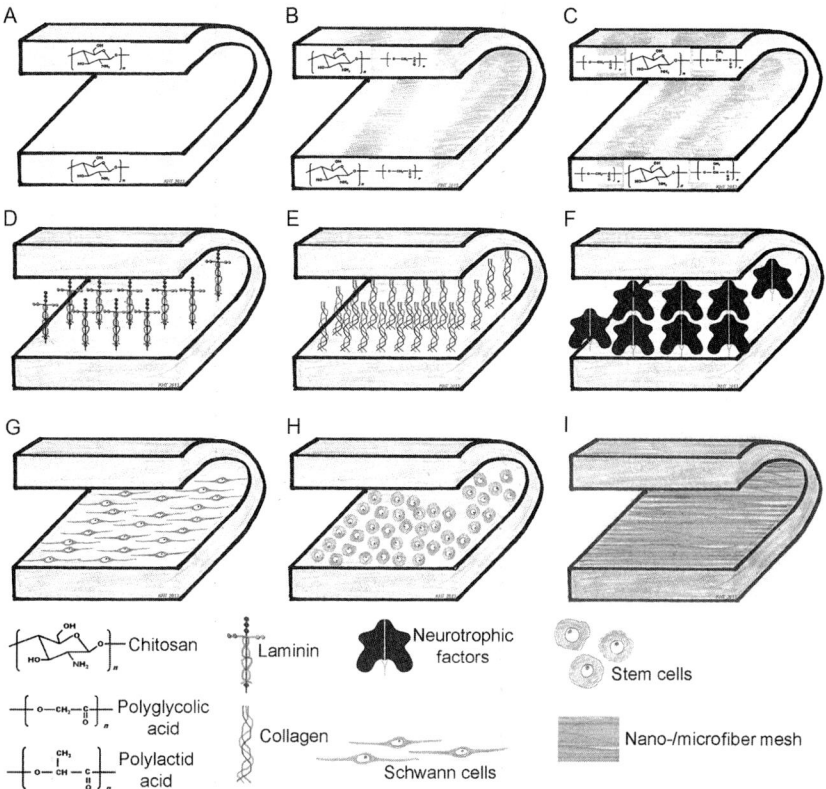

Figure 1.1 Summary of the most common chitosan-blend materials used for the fabrication of nerve graft devices. In general, basic materials are composed of (A) chitosan in different degrees of acetylation to tailor their rate of biodegradability or chitosan-blended with, for example, polyglycolic acid (PGA) (B), polylactid acid (PLA), or other commonly used polymers. Also, blends of chitosan with more than one polymer have been described, for example, PGA and PLA (C). Chitosan-based materials can further be modified at their surfaces which will be in contact to the regenerating neural tissue: extracellular matrix glycoproteins-like laminin (D) or laminin-derived peptides can be added as an collagen (E) or diverse neurotrophic factors (F), like NGF, GDNF, CNTF, or NT-3. The inner surfaces of the guidance channels can be manufactured to allow seeding of regeneration-supporting cell types, like aligned Schwann cells (G) or different types of stem cells (H). Finally, the innovative electrospinning technique can be utilized to secondary structure the surface of the chitosan-based materials by adding oriented nano- or microfibers (I). (For color version of this figure, the reader is referred to the online version of this chapter.)

neurons and glial cells. Yet, experimental results obtained from several animal models *in vivo* and first data from the first few clinical studies which were done so far have shown that chitosan–based scaffolds are good candidates for developing innovative devices for neural repair of both the CNS and PNS.

Chitosan is biocompatible, biodegradable, and its chemicophysical properties can be easily manipulated with the goal to create scaffolds with different structural features (i.e., biodegradation time or surface properties). Figure 1.1 summarizes the most cited chitosan and chitosan-blend nerve conduits (Fig. 1.1A–C) as well as the most frequently investigated surface modifications of chitosan-based nerve conduits (Fig. 1.1D–I).

After reviewing the literature regarding the use of chitosan in neural repair approaches, it can be foreseen that the time for clinical trials utilizing chitosan-based nerve regeneration-promoting devices is approaching quickly.

ACKNOWLEDGMENTS

The research leading to this chapter has received funding from the European Community's Seventh Framework Programme (FP7-HEALTH-2011) under grant agreement no. 278612 (BIOHYBRID).

REFERENCES

Acosta, N., Aranaz, I., Peniche, C., & Heras, A. (2003). Tramadol release from a delivery system based on alginate–chitosan microcapsules. *Macromolecular Bioscience, 3*, 546–551.

Aiba, S. (1992). Studies on chitosan: 4. Lysozymic hydrolysis of partially N-acetylated chitosans. *International Journal of Biological Macromolecules, 14*(4), 225–228.

Amado, S., Simoes, M. J., Armada da Silva, P. A., Luis, A. L., Shirosaki, Y., Lopes, M. A., et al. (2008). Use of hybrid chitosan membranes and N1E-115 cells for promoting nerve regeneration in an axonotmesis rat model. *Biomaterials, 29*(33), 4409–4419. http://dx. doi.org/10.1016/j.biomaterials.2008.07.043.

Anderson, J. M., Rodriguez, A., & Chang, D. T. (2008). Foreign body reaction to biomaterials. *Seminars in Immunology, 20*(2), 86–100. http://dx.doi.org/10.1016/j. smim.2007.11.004.

Ao, Q., Fung, C. K., Tsui, A. Y., Cai, S., Zuo, H. C., Chan, Y. S., et al. (2011). The regeneration of transected sciatic nerves of adult rats using chitosan nerve conduits seeded with bone marrow stromal cell-derived Schwann cells. *Biomaterials, 32*(3), 787–796. http:// dx.doi.org/10.1016/j.biomaterials.2010.09.046.

Ao, Q., Wang, A., Cao, W., Zhang, L., Kong, L., He, Q., et al. (2006). Manufacture of multimicrotubule chitosan nerve conduits with novel molds and characterization in vitro. *Journal of Biomedical Materials Research. Part A, 77*(1), 11–18. http://dx.doi.org/10.1002/jbm. a.30593.

Aranaz, I., Mengibar, M., Harris, R., Panos, I., Miralles, B., Acosta, N., et al. (2009). Functiona characterization of chitin and chitosan. *Current Chemical Biology, 3*, 203–230.

ASTM, (2001). *F2103-01 Standard guide for characterization and testing of chitosan salts as starting materials intended for use in biomedical and tissue-engineered medical product applications.*

Baulieu, E., & Schumacher, M. (2000). Progesterone as a neuroactive neurosteroid, with special reference to the effect of progesterone on myelination. *Steroids, 65*(10–11), 605–612.

Bozkurt, G., Mothe, A. J., Zahir, T., Kim, H., Shoichet, M. S., & Tator, C. H. (2010). Chitosan channels containing spinal cord-derived stem/progenitor cells for repair of subacute spinal cord injury in the rat. *Neurosurgery, 67*(6), 1733–1744. http://dx.doi.org/ 10.1227/NEU.0b013e3181f9af35.

Bray, G. M., Villegas-Perez, M. P., Vidal-Sanz, M., & Aguayo, A. J. (1987). The use of peripheral nerve grafts to enhance neuronal survival, promote growth and permit terminal reconnections in the central nervous system of adult rats. *The Journal of Experimental Biology, 132,* 5–19.

Bunge, M. B. (2002). Bridging the transected or contused adult rat spinal cord with Schwann cell and olfactory ensheathing glia transplants. *Progress in Brain Research, 137,* 275–282.

Cao, W., Cheng, M., Ao, Q., Gong, Y., Zhao, N., & Zhang, X. (2005). Physical, mechanical and degradation properties, and schwann cell affinity of cross-linked chitosan films. *Journal of Biomaterials Science. Polymer Edition, 16*(6), 791–807.

Chavez-Delgado, M. E., Gomez-Pinedo, U., Feria-Velasco, A., Huerta-Viera, M., Castaneda, S. C., Toral, F. A., et al. (2005). Ultrastructural analysis of guided nerve regeneration using progesterone- and pregnenolone-loaded chitosan prostheses. *Journal of Biomedical Materials Research. Part B, Applied Biomaterials, 74*(1), 589–600. http://dx.doi.org/10.1002/jbm.b.30243.

Chavez-Delgado, M. E., Mora-Galindo, J., Gomez-Pinedo, U., Feria-Velasco, A., Castro-Castaneda, S., Lopez-Dellamary Toral, F. A., et al. (2003). Facial nerve regeneration through progesterone-loaded chitosan prosthesis. A preliminary report. *Journal of Biomedical Materials Research Part B: Applied Biomaterials, 67*(2), 702–711. http://dx.doi.org/10.1002/jbm.b.10059.

Chen, X., Yang, Y., Yao, J., Lin, W., Li, Y., Chen, Y., et al. (2011). Bone marrow stromal cells-loaded chitosan conduits promote repair of complete transection injury in rat spinal cord. *Journal of Materials Science. Materials in Medicine, 22*(10), 2347–2356. http://dx.doi.org/10.1007/s10856-011-4401-9.

Cheng, M., Cao, W., Gao, Y., Gong, Y., Zhao, N., & Zhang, X. (2003). Studies on nerve cell affinity of biodegradable modified chitosan films. *Journal of Biomaterials Science. Polymer Edition, 14*(10), 1155–1167.

Cheng, H., Cao, Y., & Olson, L. (1996). Spinal cord repair in adult paraplegic rats: Partial restoration of hind limb function. *Science, 273*(5274), 510–513.

Cheng, M., Deng, J., Yang, F., Gong, Y., Zhao, N., & Zhang, X. (2003). Study on physical properties and nerve cell affinity of composite films from chitosan and gelatin solutions. *Biomaterials, 24*(17), 2871–2880.

Cheng, H., Huang, Y. C., Chang, P. T., & Huang, Y. Y. (2007). Laminin-incorporated nerve conduits made by plasma treatment for repairing spinal cord injury. *Biochemical and Biophysical Research Communications, 357*(4), 938–944. http://dx.doi.org/10.1016/j.bbrc.2007.04.049.

Chiono, V., Pulieri, E., Vozzi, G., Ciardelli, G., Ahluwalia, A., & Giusti, P. (2008). Genipin-crosslinked chitosan/gelatin blends for biomedical applications. *Journal of Materials Science. Materials in Medicine, 19*(2), 889–898. http://dx.doi.org/10.1007/s10856-007-3212-5.

Cho, Y., Shi, R., & Borgens, R. B. (2010). Chitosan produces potent neuroprotection and physiological recovery following traumatic spinal cord injury. *The Journal of Experimental Biology, 213*(Pt. 9), 1513–1520. http://dx.doi.org/10.1242/jeb.035162.

Crompton, K. E., Goud, J. D., Bellamkonda, R. V., Gengenbach, T. R., Finkelstein, D. I., Horne, M. K., et al. (2007). Polylysine-functionalised thermoresponsive chitosan hydrogel for neural tissue engineering. *Biomaterials, 28*(3), 441–449. http://dx.doi.org/10.1016/j.biomaterials.2006.08.044.

Cunha-Reis, C., TuzlaKoglu, K., Baas, E., Yang, Y., El Haj, A., & Reis, R. L. (2007). Influence of porosity and fibre diameter on the degradation of chitosan fibre-mesh scaffolds and cell adhesion. *Journal of Materials Science. Materials in Medicine, 18*(2), 195–200. http://dx.doi.org/10.1007/s10856-006-0681-x.

Desarnaud, F., Do Thi, A. N., Brown, A. M., Lemke, G., Suter, U., Baulieu, E. E., et al. (1998). Progesterone stimulates the activity of the promoters of peripheral myelin protein-22 and protein zero genes in Schwann cells. *Journal of Neurochemistry*, *71*(4), 1765–1768.

Dhiman, H. K., Ray, A. R., & Panda, A. K. (2004). Characterization and evaluation of chitosan matrix for in vitro growth of MCF-7 breast cancer cell lines. *Biomaterials*, *25*(21), 5147–5154. http://dx.doi.org/10.1016/j.biomaterials.2003.12.025.

Di Vita, G., Patti, R., Sparacello, M., Balistreri, C. R., Candore, G., & Caruso, C. (2008). Impact of different texture of polypropylene mesh on the inflammatory response. *International Journal of Immunopathology and Pharmacology*, *21*(1), 207–214.

Ding, F., Wu, J., Yang, Y., Hu, W., Zhu, Q., Tang, X., et al. (2010). Use of tissue-engineered nerve grafts consisting of a chitosan/poly(lactic-co-glycolic acid)-based scaffold included with bone marrow mesenchymal cells for bridging 50-mm dog sciatic nerve gaps. *Tissue Engineering. Part A*, *16*(12), 3779–3790. http://dx.doi.org/10.1089/ten.TEA.2010.0299.

Domard, A., & Domard, M. (2002). Chitosan: Structure-properties relationship and biomedical applications. In S. Dumitriu (Ed.), *Polymeric, biomaterials* (pp. 187–212). New York: Marcel Dekker, Inc.

Drury, J. L., & Mooney, D. J. (2003). Hydrogels for tissue engineering: Scaffold design variables and applications. *Biomaterials*, *24*(24), 4337–4351.

Eroglu, H., Nemutlu, E., Turkoglu, O. F., Nacar, O., Bodur, E., Sargon, M. F., et al. (2010). A quadruped study on chitosan microspheres containing atorvastatin calcium: Preparation, characterization, quantification and in-vivo application. *Chemical & Pharmaceutical Bulletin*, *58*(9), 1161–1167.

European Pharmacopeia, (2002). 1774–1775.

Evans, G. R., Brandt, K., Widmer, M. S., Lu, L., Meszlenyi, R. K., Gupta, P. K., et al. (1999). In vivo evaluation of poly(L-lactic acid) porous conduits for peripheral nerve regeneration. *Biomaterials*, *20*(12), 1109–1115.

Fan, W., Gu, J., Hu, W., Deng, A., Ma, Y., Liu, J., et al. (2008). Repairing a 35-mm-long median nerve defect with a chitosan/PGA artificial nerve graft in the human: A case study. *Microsurgery*, *28*(4), 238–242. http://dx.doi.org/10.1002/micr.20488.

Fang, P., Gao, Q., Liu, W. J., Qi, X. X., Li, G. B., Zhang, J., et al. (2010). Survival and differentiation of neuroepithelial stem cells on chitosan bicomponent fibers. *The Chinese Journal of Physiology*, *53*(4), 208–214.

Foster, L. J., & Karsten, E. (2012). A chitosan based, laser activated thin film surgical adhesive, 'SurgiLux': Preparation and demonstration. *Journal of Visualized Experiments*, *68*, 1–7. http://dx.doi.org/10.3791/3527.

Freier, T., Koh, H. S., Kazazian, K., & Shoichet, M. S. (2005). Controlling cell adhesion and degradation of chitosan films by N-acetylation. *Biomaterials*, *26*(29), 5872–5878. http://dx.doi.org/10.1016/j.biomaterials.2005.02.033.

Freier, T., Montenegro, R., Shan Koh, H., & Shoichet, M. S. (2005). Chitin-based tubes for tissue engineering in the nervous system. *Biomaterials*, *26*(22), 4624–4632. http://dx.doi.org/10.1016/j.biomaterials.2004.11.040.

Gan, Q., & Wang, T. (2007). Chitosan nanoparticle as protein delivery carrier—Systematic examination of fabrication conditions for efficient loading and release. *Colloids and Surfaces. B, Biointerfaces*, *59*(1), 24–34. http://dx.doi.org/10.1016/j.colsurfb.2007.04.009.

Gazzeri, R., Neroni, M., Alfieri, A., Galarza, M., Faiola, A., Esposito, S., et al. (2009). Transparent equine collagen biomatrix as dural repair. A prospective clinical study. *Acta Neurochirurgica*, *151*(5), 537–543. http://dx.doi.org/10.1007/s00701-009-0290-9.

Ghanaati, S., Barbeck, M., Orth, C., Willershausen, I., Thimm, B. W., Hoffmann, C., et al. (2010). Influence of beta-tricalcium phosphate granule size and morphology on tissue reaction in vivo. *Acta Biomaterialia*, *6*(12), 4476–4487. http://dx.doi.org/10.1016/j.actbio.2010.07.006.

Gong, Y., Gong, L., Gu, X., & Ding, F. (2009). Chitooligosaccharides promote peripheral nerve regeneration in a rabbit common peroneal nerve crush injury model. *Microsurgery, 29*(8), 650–656. http://dx.doi.org/10.1002/micr.20686.

Goraltchouk, A., Scanga, V., Morshead, C. M., & Shoichet, M. S. (2006). Incorporation of protein-eluting microspheres into biodegradable nerve guidance channels for controlled release. *Journal of Controlled Release, 110*(2), 400–407. http://dx.doi.org/10.1016/j.jconrel.2005.10.019.

Graf, J., Iwamoto, Y., Sasaki, M., Martin, G. R., Kleinman, H. K., Robey, F. A., et al. (1987). Identification of an amino acid sequence in laminin mediating cell attachment, chemotaxis, and receptor binding. *Cell, 48*(6), 989–996.

Gravvanis, A. I., Tsoutsos, D. A., Tagaris, G. A., Papalois, A. E., Patralexis, C. G., Iconomou, T. G., et al. (2004). Beneficial effect of nerve growth factor-7S on peripheral nerve regeneration through inside-out vein grafts: An experimental study. *Microsurgery, 24*(5), 408–415. http://dx.doi.org/10.1002/micr.20055.

Grothe, C., Haastert, K., & Jungnickel, J. (2006). Physiological function and putative therapeutic impact of the FGF-2 system in peripheral nerve regeneration—Lessons from in vivo studies in mice and rats. *Brain Research Reviews, 51*(2), 293–299. http://dx.doi.org/10.1016/j.brainresrev.2005.12.001.

Gu, J., Hu, W., Deng, A., Zhao, Q., Lu, S., & Gu, X. (2012). Surgical repair of a 30 mm long human median nerve defect in the distal forearm by implantation of a chitosan-PGA nerve guidance conduit. *Journal of Tissue Engineering and Regenerative Medicine, 6*(2), 163–168. http://dx.doi.org/10.1002/term.407.

Guenard, V., Kleitman, N., Morrissey, T. K., Bunge, R. P., & Aebischer, P. (1992). Syngeneic Schwann cells derived from adult nerves seeded in semipermeable guidance channels enhance peripheral nerve regeneration. *The Journal of Neuroscience, 12*(9), 3310–3320.

Guennoun, R., Benmessahel, Y., Delespierre, B., Gouezou, M., Rajkowski, K. M., Baulieu, E. E., et al. (2001). Progesterone stimulates Krox-20 gene expression in Schwann cells. *Brain Research. Molecular Brain Research, 90*(1), 75–82.

Guo, X., Zahir, T., Mothe, A., Shoichet, M. S., Morshead, C. M., Katayama, Y., et al. (2012). The effect of growth factors and soluble Nogo-66 receptor protein on transplanted neural stem/progenitor survival and axonal regeneration after complete transection of rat spinal cord. *Cell Transplantation, 21*(6), 1177–1197. http://dx.doi.org/10.3727/096368911X612503.

Gustafson, S. B., Fulkerson, P., Bildfell, R., Aguilera, L., & Hazzard, T. M. (2007). Chitosan dressing provides hemostasis in swine femoral arterial injury model. *Prehospital Emergency Care, 11*(2), 172–178. http://dx.doi.org/10.1080/10903120701205893.

Haipeng, G., Yinghui, Z., Jianchun, L., Yandao, G., Nanming, Z., & Xiufang, Z. (2000). Studies on nerve cell affinity of chitosan-derived materials. *Journal of Biomedical Materials Research, 52*(2), 285–295.

Han, H., Ao, Q., Chen, G., Wang, S., & Zuo, H. (2010). A novel basic fibroblast growth factor delivery system fabricated with heparin-incorporated fibrin-fibronectin matrices for repairing rat sciatic nerve disruptions. *Biotechnology Letters, 32*(4), 585–591. http://dx.doi.org/10.1007/s10529-009-0186-z.

Heath, C. A., & Rutkowski, G. E. (1998). The development of bioartificial nerve grafts for peripheral-nerve regeneration. *Trends in Biotechnology, 16*(4), 163–168.

Hirano, S., Ohe, Y., & Ono, H. (1976). Selective N-acylation of chitosan. *Carbohydrate Research, 47*(2), 315–320.

Hirano, S., Tsuchida, H., & Nagao, N. (1989). N-acetylation in chitosan and the rate of its enzymic hydrolysis. *Biomaterials, 10*(8), 574–576.

Ho, M. H., Wang, D. M., Hsieh, H. J., Liu, H. C., Hsien, T. Y., Lai, J. Y., et al. (2005). Preparation and characterization of RGD-immobilized chitosan scaffolds. *Biomaterials*, *26*(16), 3197–3206. http://dx.doi.org/10.1016/j.biomaterials.2004.08.032.

Hsu, S. H., Kuo, W. C., Chen, Y. T., Yen, C. T., Chen, Y. F., Chen, K. S., et al. (2013). New nerve regeneration strategy combining laminin-coated chitosan conduits and stem cell therapy. *Acta Biomaterialia*, *9*(5), 6606–6615. http://dx.doi.org/10.1016/j.actbio.2013.01.025.

Hsu, S. H., Lu, P. S., Ni, H. C., & Su, C. H. (2007). Fabrication and evaluation of microgrooved polymers as peripheral nerve conduits. *Biomedical Microdevices*, *9*(5), 665–674. http://dx.doi.org/10.1007/s10544-007-9068-0.

Hsueh, Y. Y., Chiang, Y. L., Wu, C. C., & Lin, S. C. (2012). Spheroid formation and neural induction in human adipose-derived stem cells on a chitosan-coated surface. *Cells, Tissues, Organs*, *196*(2), 117–128. http://dx.doi.org/10.1159/000332045.

Hu, X., Huang, J., Ye, Z., Xia, L., Li, M., Lv, B., et al. (2009). A novel scaffold with longitudinally oriented microchannels promotes peripheral nerve regeneration. *Tissue Engineering. Part A*, *15*(11), 3297–3308. http://dx.doi.org/10.1089/ten.TEA.2009.0017.

Hu, N., Wu, H., Xue, C., Gong, Y., Wu, J., Xiao, Z., et al. (2013). Long-term outcome of the repair of 50 mm long median nerve defects in rhesus monkeys with marrow mesenchymal stem cells-containing, chitosan-based tissue engineered nerve grafts. *Biomaterials*, *34*(1), 100–111. http://dx.doi.org/10.1016/j.biomaterials.2012.09.020.

Huang, Y. C., Hsu, S. H., Kuo, W. C., Chang-Chien, C. L., Cheng, H., & Huang, Y. Y. (2011). Effects of laminin-coated carbon nanotube/chitosan fibers on guided neurite growth. *Journal of Biomedical Materials Research. Part A*, *99*(1), 86–93. http://dx.doi.org/10.1002/jbm.a.33164.

Huang, J., Hu, X., Lu, L., Ye, Z., Zhang, Q., & Luo, Z. (2010). Electrical regulation of Schwann cells using conductive polypyrrole/chitosan polymers. *Journal of Biomedical Materials Research. Part A*, *93*(1), 164–174. http://dx.doi.org/10.1002/jbm.a.32511.

Huang, Y. C., Huang, C. C., Huang, Y. Y., & Chen, K. S. (2007). Surface modification and characterization of chitosan or PLGA membrane with laminin by chemical and oxygen plasma treatment for neural regeneration. *Journal of Biomedical Materials Research. Part A*, *82*(4), 842–851. http://dx.doi.org/10.1002/jbm.a.31036.

Huang, Y. C., Huang, Y. Y., Huang, C. C., & Liu, H. C. (2005). Manufacture of porous polymer nerve conduits through a lyophilizing and wire-heating process. *Journal of Biomedical Materials Research. Part B, Applied Biomaterials*, *74*(1), 659–664. http://dx.doi.org/10.1002/jbm.b.30267.

Huang, M., Khor, E., & Lim, L. Y. (2004). Uptake and cytotoxicity of chitosan molecules and nanoparticles: Effects of molecular weight and degree of deacetylation. *Pharmaceutical Research*, *21*(2), 344–353.

Huang, J., Lu, L., Hu, X., Ye, Z., Peng, Y., Yan, X., et al. (2010). Electrical stimulation accelerates motor functional recovery in the rat model of 15-mm sciatic nerve gap bridged by scaffolds with longitudinally oriented microchannels. *Neurorehabilitation and Neural Repair*, *24*(8), 736–745. http://dx.doi.org/10.1177/1545968310368686.

Huang, Y., Onyeri, S., Siewe, M., Moshfeghian, A., & Madihally, S. V. (2005). In vitro characterization of chitosan-gelatin scaffolds for tissue engineering. *Biomaterials*, *26*(36), 7616–7627. http://dx.doi.org/10.1016/j.biomaterials.2005.05.036.

Ide, C. (1996). Peripheral nerve regeneration. *Neuroscience Research*, *25*(2), 101–121.

Ishikawa, N., Suzuki, Y., Ohta, M., Cho, H., Suzuki, S., Dezawa, M., et al. (2007). Peripheral nerve regeneration through the space formed by a chitosan gel sponge. *Journal of Biomedical Materials Research. Part A*, *83*(1), 33–40. http://dx.doi.org/10.1002/jbm.a.31126.

Itoh, S., Matsuda, A., Kobayashi, H., Ichinose, S., Shinomiya, K., & Tanaka, J. (2005). Effects of a laminin peptide (YIGSR) immobilized on crab-tendon chitosan tubes on nerve regeneration. *Journal of Biomedical Materials Research. Part B, Applied Biomaterials*, *73*(2), 375–382. http://dx.doi.org/10.1002/jbm.b.30224.

Itoh, S., Suzuki, M., Yamaguchi, I., Takakuda, K., Kobayashi, H., Shinomiya, K., et al. (2003). Development of a nerve scaffold using a tendon chitosan tube. *Artificial Organs*, *27*(12), 1079–1088.

Itoh, S., Yamaguchi, I., Suzuki, M., Ichinose, S., Takakuda, K., Kobayashi, H., et al. (2003). Hydroxyapatite-coated tendon chitosan tubes with adsorbed laminin peptides facilitate nerve regeneration in vivo. *Brain Research*, *993*(1–2), 111–123.

Itosaka, H., Kuroda, S., Shichinohe, H., Yasuda, H., Yano, S., Kamei, S., et al. (2009). Fibrin matrix provides a suitable scaffold for bone marrow stromal cells transplanted into injured spinal cord: A novel material for CNS tissue engineering. *Neuropathology*, *29*(3), 248–257. http://dx.doi.org/10.1111/j.1440-1789.2008.00971.x.

Janes, K. A., Fresneau, M. P., Marazuela, A., Fabra, A., & Alonso, M. J. (2001). Chitosan nanoparticles as delivery systems for doxorubicin. *Journal of Controlled Release*, *73*(2–3), 255–267.

Jiang, M., Zhuge, X., Yang, Y., Gu, X., & Ding, F. (2009). The promotion of peripheral nerve regeneration by chitooligosaccharides in the rat nerve crush injury model. *Neuroscience Letters*, *454*(3), 239–243. http://dx.doi.org/10.1016/j.neulet.2009.03.042.

Jiao, H., Yao, J., Yang, Y., Chen, X., Lin, W., Li, Y., et al. (2009). Chitosan/polyglycolic acid nerve grafts for axon regeneration from prolonged axotomized neurons to chronically denervated segments. *Biomaterials*, *30*(28), 5004–5018. http://dx.doi.org/10.1016/j.biomaterials.2009.05.059.

Johnson, P. J., Parker, S. R., & Sakiyama-Elbert, S. E. (2009). Controlled release of neurotrophin-3 from fibrin-based tissue engineering scaffolds enhances neural fiber sprouting following subacute spinal cord injury. *Biotechnology and Bioengineering*, *104*(6), 1207–1214. http://dx.doi.org/10.1002/bit.22476.

Jung-Testas, I., Schumacher, M., Robel, P., & Baulieu, E. E. (1996). Demonstration of progesterone receptors in rat Schwann cells. *The Journal of Steroid Biochemistry and Molecular Biology*, *58*(1), 77–82.

Kam, H. M., Khor, E., & Lim, L. Y. (1999). Storage of partially deacetylated chitosan films. *Journal of Biomedical Materials Research*, *48*(6), 881–888.

Kang, C. E., Poon, P. C., Tator, C. H., & Shoichet, M. S. (2009). A new paradigm for local and sustained release of therapeutic molecules to the injured spinal cord for neuroprotection and tissue repair. *Tissue Engineering. Part A*, *15*(3), 595–604. http://dx.doi.org/10.1089/ten.tea.2007.0349.

Khor, E., & Lim, L. Y. (2003). Implantable applications of chitin and chitosan. *Biomaterials*, *24*(13), 2339–2349.

Kim, H., Tator, C. H., & Shoichet, M. S. (2008). Design of protein-releasing chitosan channels. *Biotechnology Progress*, *24*(4), 932–937. http://dx.doi.org/10.1021/bp070352a.

Kim, H., Zahir, T., Tator, C. H., & Shoichet, M. S. (2011). Effects of dibutyryl cyclic-AMP on survival and neuronal differentiation of neural stem/progenitor cells transplanted into spinal cord injured rats. *PLoS One*, *6*(6), e21744. http://dx.doi.org/10.1371/journal.pone.0021744.

Kleinman, H. K., Ogle, R. C., Cannon, F. B., Little, C. D., Sweeney, T. M., & Luckenbill-Edds, L. (1988). Laminin receptors for neurite formation. *Proceedings of the National Academy of Sciences of the United States of America*, *85*(4), 1282–1286.

Koenig, H. L., Gong, W. H., & Pelissier, P. (2000). Role of progesterone in peripheral nerve repair. *Reviews of Reproduction*, *5*(3), 189–199.

Kofuji, K., Ito, T., Murata, Y., & Kawashima, S. (2001). Biodegradation and drug release of chitosan gel beads in subcutaneous air pouches of mice. *Biological & Pharmaceutical Bulletin*, *24*(2), 205–208.

Kohane, D. S., Tse, J. Y., Yeo, Y., Padera, R., Shubina, M., & Langer, R. (2006). Biodegradable polymeric microspheres and nanospheres for drug delivery in the peritoneum. *Journal of Biomedical Materials Research. Part A*, 77(2), 351–361. http://dx.doi.org/10.1002/jbm.a.30654.

Kumar, M. N. V. (2002). Chitosan: Structure-properties relation-ship and biomedical applications. In S. Dumitriu (Ed.), *Polymeric, biomaterials* (pp. 187–212). New York: Marcel Dekker, Inc.

Kumar, M. N., Muzzarelli, R. A., Muzzarelli, C., Sashiwa, H., & Domb, A. J. (2004). Chitosan chemistry and pharmaceutical perspectives. *Chemical Reviews*, 104(12), 6017–6084. http://dx.doi.org/10.1021/cr030441b.

Kuo, Y. C., & Lin, C. C. (2013). Accelerated nerve regeneration using induced pluripotent stem cells in chitin-chitosan-gelatin scaffolds with inverted colloidal crystal geometry. *Colloids and Surfaces. B, Biointerfaces*, 103, 595–600. http://dx.doi.org/10.1016/j.colsurfb.2012.11.001.

Kurita, K., Kaji, Y., Mori, T., & Nishiyama, Y. (2000). Enzymatic degradation of beta-chitin: Susceptibility and the influence of deacetylation. *Carbohydrate Polymers*, 42(1), 19–21.

Lauto, A., Foster, L. J., Avolio, A., Sampson, D., Raston, C., Sarris, M., et al. (2008). Sutureless nerve repair with laser-activated chitosan adhesive: A pilot in vivo study. *Photomedicine and Laser Surgery*, 26(3), 227–234. http://dx.doi.org/10.1089/pho.2007.2131.

Lauto, A., Stoodley, M., Marcel, H., Avolio, A., Sarris, M., McKenzie, G., et al. (2007). In vitro and in vivo tissue repair with laser-activated chitosan adhesive. *Lasers in Surgery and Medicine*, 39(1), 19–27. http://dx.doi.org/10.1002/lsm.20418.

Leipzig, N. D., Wylie, R. G., Kim, H., & Shoichet, M. S. (2011). Differentiation of neural stem cells in three-dimensional growth factor-immobilized chitosan hydrogel scaffolds. *Biomaterials*, 32(1), 57–64. http://dx.doi.org/10.1016/j.biomaterials.2010.09.031.

Lemmon, V., Burden, S. M., Payne, H. R., Elmslie, G. J., & Hlavin, M. L. (1992). Neurite growth on different substrates: Permissive versus instructive influences and the role of adhesive strength. *The Journal of Neuroscience*, 12(3), 818–826.

Li, Y., Field, P. M., & Raisman, G. (1997). Repair of adult rat corticospinal tract by transplants of olfactory ensheathing cells. *Science*, 277(5334), 2000–2002.

Li, G. N., & Hoffman-Kim, D. (2008). Tissue-engineered platforms of axon guidance. *Tissue Engineering. Part B, Reviews*, 14(1), 33–51. http://dx.doi.org/10.1089/teb.2007.0181.

Li, X., Wang, W., Wei, G., Wang, G., Zhang, W., & Ma, X. (2010). Immunophilin FK506 loaded in chitosan guide promotes peripheral nerve regeneration. *Biotechnology Letters*, 32(9), 1333–1337. http://dx.doi.org/10.1007/s10529-010-0287-8.

Li, X., Yang, Z., Zhang, A., Wang, T., & Chen, W. (2009). Repair of thoracic spinal cord injury by chitosan tube implantation in adult rats. *Biomaterials*, 30(6), 1121–1132. http://dx.doi.org/10.1016/j.biomaterials.2008.10.063.

Lim, L. Y., Khor, E., & Ling, C. E. (1999). Effects of dry heat and saturated steam on the physical properties of chitosan. *Journal of Biomedical Materials Research*, 48(2), 111–116.

Lin, Y. L., Jen, J. C., Hsu, S. H., & Chiu, I. M. (2008). Sciatic nerve repair by microgrooved nerve conduits made of chitosan-gold nanocomposites. *Surgical Neurology*, 70(Suppl. 1), 9–18. http://dx.doi.org/10.1016/j.surneu.2008.01.057.

Liu, Y., Hou, C., Lin, H., & Wei, C. (2011). An experimental study on effect of chitosan/polyvinyl alcohol nerve conduits on peripheral nerve regeneration in macaques. *Zhongguo Xiu Fu Chong Jian Wai Ke Za Zhi*, 25(10), 1235–1238.

Ma, L., Gao, C., Mao, Z., Zhou, J., Shen, J., Hu, X., et al. (2003). Collagen/chitosan porous scaffolds with improved biostability for skin tissue engineering. *Biomaterials*, 24(26), 4833–4841.

Madduri, S., Feldman, K., Tervoort, T., Papaloizos, M., & Gander, B. (2010). Collagen nerve conduits releasing the neurotrophic factors GDNF and NGF. *Journal of Controlled Release*, 143(2), 168–174. http://dx.doi.org/10.1016/j.jconrel.2009.12.017.

Madduri, S., Papaloizos, M., & Gander, B. (2010). Trophically and topographically functionalized silk fibroin nerve conduits for guided peripheral nerve regeneration. *Biomaterials, 31*(8), 2323–2334. http://dx.doi.org/10.1016/j.biomaterials.2009.11.073.

Madihally, S. V., & Matthew, H. W. (1999). Porous chitosan scaffolds for tissue engineering. *Biomaterials, 20*(12), 1133–1142.

Madison, R., da Silva, C. F., Dikkes, P., Chiu, T. H., & Sidman, R. L. (1985). Increased rate of peripheral nerve regeneration using bioresorbable nerve guides and a laminin-containing gel. *Experimental Neurology, 88*(3), 767–772.

Madison, R. D., da Silva, C., Dikkes, P., Sidman, R. L., & Chiu, T. H. (1987). Peripheral nerve regeneration with entubulation repair: Comparison of biodegradeable nerve guides versus polyethylene tubes and the effects of a laminin-containing gel. *Experimental Neurology, 95*(2), 378–390.

Manthorpe, M., Engvall, E., Ruoslahti, E., Longo, F. M., Davis, G. E., & Varon, S. (1983). Laminin promotes neuritic regeneration from cultured peripheral and central neurons. *The Journal of Cell Biology, 97*(6), 1882–1890.

Marcol, W., Larysz-Brysz, M., Kucharska, M., Niekraszewicz, A., Slusarczyk, W., Kotulska, K., et al. (2011). Reduction of post-traumatic neuroma and epineural scar formation in rat sciatic nerve by application of microcrystallic chitosan. *Microsurgery, 31*(8), 642–649. http://dx.doi.org/10.1002/micr.20945.

Marreco, P. R., da Luz Moreira, P., Genari, S. C., & Moraes, A. M. (2004). Effects of different sterilization methods on the morphology, mechanical properties, and cytotoxicity of chitosan membranes used as wound dressings. *Journal of Biomedical Materials Research. Part B, Applied Biomaterials, 71*(2), 268–277. http://dx.doi.org/10.1002/jbm.b.30081.

Matsuda, A., Kobayashi, H., Itoh, S., Kataoka, K., & Tanaka, J. (2005). Immobilization of laminin peptide in molecularly aligned chitosan by covalent bonding. *Biomaterials, 26*(15), 2273–2279. http://dx.doi.org/10.1016/j.biomaterials.2004.07.032.

Matsumoto, I., Kaneko, M., Oda, M., & Watanabe, G. (2010). Repair of intra-thoracic autonomic nerves using chitosan tubes. *Interactive Cardiovascular and Thoracic Surgery, 10*(4), 498–501. http://dx.doi.org/10.1510/icvts.2009.227744.

Mingyu, C., Kai, G., Jiamou, L., Yandao, G., Nanming, Z., & Xiufang, Z. (2004). Surface modification and characterization of chitosan film blended with poly-L-lysine. *Journal of Biomaterials Applications, 19*(1), 59–75. http://dx.doi.org/10.1177/0885328204043450.

Mittnacht, U., Hartmann, H., Hein, S., Oliveira, H., Dong, M., Pego, A. P., et al. (2010). Chitosan/siRNA nanoparticles biofunctionalize nerve implants and enable neurite outgrowth. *Nano Letters, 10*(10), 3933–3939. http://dx.doi.org/10.1021/nl1016909.

Mumper, R., Wang, J., Claspell, J., & Rolland, A. P. (1995). Novel polimeric condensing carriers for gene delivery. In *Proceedings Interntional Symposium Controlled Release Bioactive Materials, 22,* (pp. 178–179).

Muzzarelli, R. A. A. (1977). *Chitin.* New York: Pergamon.

Nomura, H., Baladie, B., Katayama, Y., Morshead, C. M., Shoichet, M. S., & Tator, C. H. (2008). Delayed implantation of intramedullary chitosan channels containing nerve grafts promotes extensive axonal regeneration after spinal cord injury. *Neurosurgery, 63*(1), 127–141. http://dx.doi.org/10.1227/01.NEU.0000335080.47352.31, discussion 141–123.

Nomura, H., Kim, H., Mothe, A., Zahir, T., Kulbatski, I., Morshead, C. M., et al. (2010). Endogenous radial glial cells support regenerating axons after spinal cord transection. *Neuroreport, 21*(13), 871–876. http://dx.doi.org/10.1097/WNR.0b013e32833d9695.

Nomura, H., Tator, C. H., & Shoichet, M. S. (2006). Bioengineered strategies for spinal cord repair. *Journal of Neurotrauma, 23*(3–4), 496–507. http://dx.doi.org/10.1089/neu.2006.23.496.

Nomura, H., Zahir, T., Kim, H., Katayama, Y., Kulbatski, I., Morshead, C. M., et al. (2008). Extramedullary chitosan channels promote survival of transplanted neural stem and progenitor cells and create a tissue bridge after complete spinal cord transection. *Tissue Engineering. Part A, 14*(5), 649–665. http://dx.doi.org/10.1089/tea.2007.0180.

Novikova, L. N., Novikov, L. N., & Kellerth, J. O. (2003). Biopolymers and biodegradable smart implants for tissue regeneration after spinal cord injury. *Current Opinion in Neurology, 16*(6), 711–715. http://dx.doi.org/10.1097/01.wco.0000102620.38669.3e.

Novikova, L. N., Pettersson, J., Brohlin, M., Wiberg, M., & Novikov, L. N. (2008). Biodegradable poly-beta-hydroxybutyrate scaffold seeded with Schwann cells to promote spinal cord repair. *Biomaterials, 29*(9), 1198–1206. http://dx.doi.org/10.1016/j.biomaterials.2007.11.033.

Oh, S. H., Kim, J. H., Song, K. S., Jeon, B. H., Yoon, J. H., Seo, T. B., et al. (2008). Peripheral nerve regeneration within an asymmetrically porous PLGA/Pluronic F127 nerve guide conduit. *Biomaterials, 29*(11), 1601–1609. http://dx.doi.org/10.1016/j.biomaterials.2007.11.036.

Pangburn, S. H., Trescony, P. V., & Heller, J. (1982). Lysozyme degradation of partially deacetylated chitin, its films and hydrogels. *Biomaterials, 3*(2), 105–108.

Patel, M., Mao, L., Wu, B., & Vandevord, P. J. (2007). GDNF-chitosan blended nerve guides: A functional study. *Journal of Tissue Engineering and Regenerative Medicine, 1*(5), 360–367. http://dx.doi.org/10.1002/term.44.

Patel, M., Mao, L., Wu, B., & VandeVord, P. (2009). GDNF blended chitosan nerve guides: An in vivo study. *Journal of Biomedical Materials Research. Part A, 90*(1), 154–165. http://dx.doi.org/10.1002/jbm.a.32072.

Patel, M., VandeVord, P. J., Matthew, H. W., DeSilva, S., Wu, B., & Wooley, P. H. (2008a). Collagen-chitosan nerve guides for peripheral nerve repair: A histomorphometric study. *Journal of Biomaterials Applications, 23*(2), 101–121. http://dx.doi.org/10.1177/0885328207084521.

Patel, M., Vandevord, P. J., Matthew, H. W., DeSilva, S., Wu, B., & Wooley, P. H. (2008b). Functional gait evaluation of collagen chitosan nerve guides for sciatic nerve repair. *Tissue Engineering. Part C, Methods, 14*(4), 365–370. http://dx.doi.org/10.1089/ten.tec.2008.0166.

Patel, M., Vandevord, P. J., Matthew, H., Wu, B., DeSilva, S., & Wooley, P. H. (2006). Video-gait analysis of functional recovery of nerve repaired with chitosan nerve guides. *Tissue Engineering, 12*(11), 3189–3199. http://dx.doi.org/10.1089/ten.2006.12.3189.

Pavinatto, F. J., Pavinatto, A., Caseli, L., Santos, D. S., Jr., Nobre, T. M., Zaniquelli, M. E., et al. (2007). Interaction of chitosan with cell membrane models at the air–water interface. *Biomacromolecules, 8*(5), 1633–1640. http://dx.doi.org/10.1021/bm0701550.

Peniche, C., Arguelles-Monal, W., Peniche, H., & Acosta, N. (2003). Chitosan: An atractive biocompatible polymer for microencapsulation. *Macromolecular Bioscience, 3*, 511–520.

Pfister, L. A., Alther, E., Papaloizos, M., Merkle, H. P., & Gander, B. (2008). Controlled nerve growth factor release from multi-ply alginate/chitosan-based nerve conduits. *European Journal of Pharmaceutics and Biopharmaceutics, 69*(2), 563–572. http://dx.doi.org/10.1016/j.ejpb.2008.01.014.

Pfister, L. A., Papaloizos, M., Merkle, H. P., & Gander, B. (2007). Hydrogel nerve conduits produced from alginate/chitosan complexes. *Journal of Biomedical Materials Research. Part A, 80*(4), 932–937. http://dx.doi.org/10.1002/jbm.a.31052.

Pierschbacher, M. D., & Ruoslahti, E. (1984). Cell attachment activity of fibronectin can be duplicated by small synthetic fragments of the molecule. *Nature, 309*(5963), 30–33.

Prabhakaran, M. P., Venugopal, J. R., Chyan, T. T., Hai, L. B., Chan, C. K., Lim, A. Y., et al. (2008). Electrospun biocomposite nanofibrous scaffolds for neural tissue engineering. *Tissue Engineering. Part A, 14*(11), 1787–1797. http://dx.doi.org/10.1089/ten.tea.2007.0393.

Ramer, M. S., Priestley, J. V., & McMahon, S. B. (2000). Functional regeneration of sensory axons into the adult spinal cord. *Nature, 403*(6767), 312–316. http://dx.doi.org/10.1038/35002084.

Ramon-Cueto, A., Plant, G. W., Avila, J., & Bunge, M. B. (1998). Long-distance axonal regeneration in the transected adult rat spinal cord is promoted by olfactory ensheathing glia transplants. *The Journal of Neuroscience, 18*(10), 3803–3815.

Rickett, T. A., Amoozgar, Z., Tuchek, C. A., Park, J., Yeo, Y., & Shi, R. (2011). Rapidly photo-cross-linkable chitosan hydrogel for peripheral neurosurgeries. *Biomacromolecules, 12*(1), 57–65. http://dx.doi.org/10.1021/bm101004r.

Rosales-Cortes, M., Peregrina-Sandoval, J., Banuelos-Pineda, J., Castellanos-Martinez, E. E., Gomez-Pinedo, U. A., & Albarran-Rodriguez, E. (2003). Regeneration of the axotomised sciatic nerve in dogs using the tubulisation technique with Chitosan biomaterial preloaded with progesterone. *Revista de Neurologia, 36*(12), 1137–1141.

Rosales-Cortes, M., Peregrina-Sandoval, J., Banuelos-Pineda, J., Sarabia-Estrada, R., Gomez-Rodiles, C. C., Albarran-Rodriguez, E., et al. (2003). Immunological study of a chitosan prosthesis in the sciatic nerve regeneration of the axotomized dog. *Journal of Biomaterials Applications, 18*(1), 15–23.

Roy, K., Mao, H. Q., Huang, S. K., & Leong, K. W. (1999). Oral gene delivery with chitosan—DNA nanoparticles generates immunologic protection in a murine model of peanut allergy. *Nature Medicine, 5*(4), 387–391. http://dx.doi.org/10.1038/7385.

Samadikuchaksaraei, A. (2007). An overview of tissue engineering approaches for management of spinal cord injuries. *Journal of Neuroengineering and Rehabilitation, 4*, 15. http://dx.doi.org/10.1186/1743-0003-4-15.

Sarasam, A., & Madihally, S. V. (2005). Characterization of chitosan-polycaprolactone blends for tissue engineering applications. *Biomaterials, 26*(27), 5500–5508. http://dx.doi.org/10.1016/j.biomaterials.2005.01.071.

Sashiwa, H., Saimoto, H., Shigemata, Y., Ogawa, R., & Tokura, S. (1991). Distribution of the acetamide group in partially deacetylated chitins. *Carbohydrate Polymers, 16*(3), 291–296.

Sephel, G. C., Burrous, B. A., & Kleinman, H. K. (1989). Laminin neural activity and binding proteins. *Developmental Neuroscience, 11*(4–5), 313–331.

Shen, H., Shen, Z. L., Zhang, P. H., Chen, N. L., Wang, Y. C., Zhang, Z. F., et al. (2010). Ciliary neurotrophic factor-coated polylactic-polyglycolic acid chitosan nerve conduit promotes peripheral nerve regeneration in canine tibial nerve defect repair. *Journal of Biomedical Materials Research. Part B, Applied Biomaterials, 95*(1), 161–170. http://dx.doi.org/10.1002/jbm.b.31696.

Simoes, M. J., Amado, S., Gartner, A., Armada-Da-Silva, P. A., Raimondo, S., Vieira, M., et al. (2010). Use of chitosan scaffolds for repairing rat sciatic nerve defects. *Italian Journal of Anatomy and Embryology, 115*(3), 190–210.

Simoes, M. J., Gartner, A., Shirosaki, Y., Gil da Costa, R. M., Cortez, P. P., Gartner, F., et al. (2011). In vitro and in vivo chitosan membranes testing for peripheral nerve reconstruction. *Acta Médica Portuguesa, 24*(1), 43–52.

Singh, D. K., & Ray, A. R. (2000). Biomedical application of chitin and chitosan and their derivatives. *Journal of Macromolecular Science: Part C: Polymer Reviews, 40*, 69–83.

Soria, J. M., Martinez Ramos, C., Salmeron Sanchez, M., Benavent, V., Campillo Fernandez, A., Gomez Ribelles, J. L., et al. (2006). Survival and differentiation of embryonic neural explants on different biomaterials. *Journal of Biomedical Materials Research. Part A, 79*(3), 495–502. http://dx.doi.org/10.1002/jbm.a.30803.

Straley, K. S., Foo, C. W., & Heilshorn, S. C. (2010). Biomaterial design strategies for the treatment of spinal cord injuries. *Journal of Neurotrauma, 27*(1), 1–19. http://dx.doi.org/10.1089/neu.2009.0948.

Suh, J. K., & Matthew, H. W. (2000). Application of chitosan-based polysaccharide biomaterials in cartilage tissue engineering: A review. *Biomaterials, 21*(24), 2589–2598.
Suzuki, M., Itoh, S., Yamaguchi, I., Takakuda, K., Kobayashi, H., Shinomiya, K., et al.
(2003). Tendon chitosan tubes covalently coupled with synthesized laminin peptides
facilitate nerve regeneration in vivo. *Journal of Neuroscience Research, 72*(5), 646–659.
http://dx.doi.org/10.1002/jnr.10589.
Tashiro, K., Sephel, G. C., Weeks, B., Sasaki, M., Martin, G. R., Kleinman, H. K., et al.
(1989). A synthetic peptide containing the IKVAV sequence from the A chain of laminin
mediates cell attachment, migration, and neurite outgrowth. *The Journal of Biological
Chemistry, 264*(27), 16174–16182.
Teng, Y. D., Lavik, E. B., Qu, X., Park, K. I., Ourednik, J., Zurakowski, D., et al. (2002).
Functional recovery following traumatic spinal cord injury mediated by a unique polymer scaffold seeded with neural stem cells. *Proceedings of the National Academy of Sciences of
the United States of America, 99*(5), 3024–3029. http://dx.doi.org/10.1073/
pnas.052678899.
Timpl, R., Rohde, H., Robey, P. G., Rennard, S. I., Foidart, J. M., & Martin, G. R. (1979).
Laminin—A glycoprotein from basement membranes. *The Journal of Biological Chemistry,
254*(19), 9933–9937.
Tomihata, K., & Ikada, Y. (1997). In vitro and in vivo degradation of films of chitin and its
deacetylated derivatives. *Biomaterials, 18*(7), 567–575.
Tysseling-Mattiace, V. M., Sahni, V., Niece, K. L., Birch, D., Czeisler, C., Fehlings, M. G.,
et al. (2008). Self-assembling nanofibers inhibit glial scar formation and promote axon
elongation after spinal cord injury. *The Journal of Neuroscience, 28*(14), 3814–3823.
http://dx.doi.org/10.1523/JNEUROSCI.0143-08.2008.
Ueno, H., Yamada, H., Tanaka, I., Kaba, N., Matsuura, M., Okumura, M., et al. (1999).
Accelerating effects of chitosan for healing at early phase of experimental open wound
in dogs. *Biomaterials, 20*(15), 1407–1414.
Vachoud, L., & Domard, A. (2001). Physicochemical properties of physical chitin hydrogels:
Modeling and relation with the mechanical properties. *Biomacromolecules, 2*(4), 1294–1300.
Vasconcelos, B. C., & Gay-Escoda, C. (2000). Facial nerve repair with expanded poly-
tetrafluoroethylene and collagen conduits: An experimental study in the rabbit. *Journal
of Oral and Maxillofacial Surgery, 58*(11), 1257–1262. http://dx.doi.org/10.1053/
joms.2000.16626.
Wang, A., Ao, Q., Cao, W., Yu, M., He, Q., Kong, L., et al. (2006). Porous chitosan tubular
scaffolds with knitted outer wall and controllable inner structure for nerve tissue engineering. *Journal of Biomedical Materials Research. Part A, 79*(1), 36–46. http://dx.doi.
org/10.1002/jbm.a.30683.
Wang, A., Ao, Q., Wei, Y., Gong, K., Liu, X., Zhao, N., et al. (2007). Physical properties
and biocompatibility of a porous chitosan-based fiber-reinforced conduit for nerve
regeneration. *Biotechnology Letters, 29*(11), 1697–1702. http://dx.doi.org/10.1007/
s10529-007-9460-0.
Wang, X., Hu, W., Cao, Y., Yao, J., Wu, J., & Gu, X. (2005). Dog sciatic nerve regeneration
across a 30-mm defect bridged by a chitosan/PGA artificial nerve graft. *Brain, 128*(Pt. 8),
1897–1910. http://dx.doi.org/10.1093/brain/awh517.
Wang, W., Itoh, S., Konno, K., Kikkawa, T., Ichinose, S., Sakai, K., et al. (2009). Effects of
Schwann cell alignment along the oriented electrospun chitosan nanofibers on nerve
regeneration. *Journal of Biomedical Materials Research. Part A, 91*(4), 994–1005. http://
dx.doi.org/10.1002/jbm.a.32329.
Wang, W., Itoh, S., Matsuda, A., Aizawa, T., Demura, M., Ichinose, S., et al. (2008).
Enhanced nerve regeneration through a bilayered chitosan tube: The effect of introduc-
tion of glycine spacer into the CYIGSR sequence. *Journal of Biomedical Materials Research.
Part A, 85*(4), 919–928. http://dx.doi.org/10.1002/jbm.a.31522.

Wang, W., Itoh, S., Matsuda, A., Ichinose, S., Shinomiya, K., Hata, Y., et al. (2008). Influences of mechanical properties and permeability on chitosan nano/microfiber mesh tubes as a scaffold for nerve regeneration. *Journal of Biomedical Materials Research. Part A, 84*(2), 557–566. http://dx.doi.org/10.1002/jbm.a.31536.

Wang, W., Itoh, S., Yamamoto, N., Okawa, A., Nagai, A., & Yamasita, K. (2010). Enhancement of nerve regeneration along a chitosan nanofiber mesh tube on which electrically polarized beta-tricalcium phosphate particles are immobilized. *Acta Biomaterialia, 6*(10), 4027–4033.

Wang, D., Wen, Y., Lan, X., & Li, H. (2010). Experimental study on bone marrow mesenchymal stem cells seeded in chitosan-alginate scaffolds for repairing spinal cord injury. *Zhongguo Xiu Fu Chong Jian Wai Ke Za Zhi, 24*(2), 190–196.

Wang, H., Zhao, Q., Zhao, W., Liu, Q., Gu, X., & Yang, Y. (2012). Repairing rat sciatic nerve injury by a nerve-growth-factor-loaded, chitosan-based nerve conduit. *Biotechnology and Applied Biochemistry, 59*(5), 388–394. http://dx.doi.org/10.1002/bab.1031.

Wei, X., Lao, J., & Gu, Y. D. (2003). Bridging peripheral nerve defect with chitosan-collagen film. *Chinese Journal of Traumatology, 6*(3), 131–134.

Wenling, C., Duohui, J., Jiamou, L., Yandao, G., Nanming, Z., & Xiufang, Z. (2005). Effects of the degree of deacetylation on the physicochemical properties and Schwann cell affinity of chitosan films. *Journal of Biomaterials Applications, 20*(2), 157–177. http://dx.doi.org/10.1177/0885328205049897.

Willerth, S. M., & Sakiyama-Elbert, S. E. (2007). Approaches to neural tissue engineering using scaffolds for drug delivery. *Advanced Drug Delivery Reviews, 59*(4–5), 325–338. http://dx.doi.org/10.1016/j.addr.2007.03.014.

Williams, D. F. (2008). On the mechanisms of biocompatibility. *Biomaterials, 29*(20), 2941–2953. http://dx.doi.org/10.1016/j.biomaterials.2008.04.023.

Woerly, S., Pinet, E., de Robertis, L., Van Diep, D., & Bousmina, M. (2001). Spinal cord repair with PHPMA hydrogel containing RGD peptides (NeuroGel). *Biomaterials, 22*(10), 1095–1111.

Xie, F., Li, Q. F., Gu, B., Liu, K., & Shen, G. X. (2008). In vitro and in vivo evaluation of a biodegradable chitosan-PLA composite peripheral nerve guide conduit material. *Microsurgery, 28*(6), 471–479. http://dx.doi.org/10.1002/micr.20514.

Xie, F., Li, Q. F., & Zhao, L. S. (2005). Study on using a new biodegradable conduit to repairing rat's peripheral nerve defect. *Zhonghua Zheng Xing Wai Ke Za Zhi, 21*(4), 295–298.

Xu, G., Nie, D. Y., Wang, W. Z., Zhang, P. H., Shen, J., Ang, B. T., et al. (2004). Optic nerve regeneration in polyglycolic acid-chitosan conduits coated with recombinant L1-Fc. *Neuroreport, 15*(14), 2167–2172.

Xu, H., Yan, Y., & Li, S. (2011). PDLLA/chondroitin sulfate/chitosan/NGF conduits for peripheral nerve regeneration. *Biomaterials, 32*(20), 4506–4516. http://dx.doi.org/10.1016/j.biomaterials.2011.02.023.

Xu, H., Yan, Y., Wan, T., & Li, S. (2009). Degradation properties of the electrostatic assembly PDLLA/CS/CHS nerve conduit. *Biomedical Materials, 4*(4), 045006. http://dx.doi.org/10.1088/1748-6041/4/4/045006.

Xu, X. M., Zhang, S. X., Li, H., Aebischer, P., & Bunge, M. B. (1999). Regrowth of axons into the distal spinal cord through a Schwann-cell-seeded mini-channel implanted into hemisected adult rat spinal cord. *The European Journal of Neuroscience, 11*(5), 1723–1740.

Xue, C., Hu, N., Gu, Y., Yang, Y., Liu, Y., Liu, J., et al. (2012). Joint use of a chitosan/PLGA scaffold and MSCs to bridge an extra large gap in dog sciatic nerve. *Neurorehabilitation and Neural Repair, 26*(1), 96–106. http://dx.doi.org/10.1177/1545968311420444.

Yamaguchi, I., Itoh, S., Suzuki, M., Osaka, A., & Tanaka, J. (2003). The chitosan prepared from crab tendons: II. The chitosan/apatite composites and their application to nerve regeneration. *Biomaterials, 24*(19), 3285–3292.

Yang, Y., Chen, X., Ding, F., Zhang, P., Liu, J., & Gu, X. (2007). Biocompatibility evaluation of silk fibroin with peripheral nerve tissues and cells in vitro. *Biomaterials, 28*(9), 1643–1652. http://dx.doi.org/10.1016/j.biomaterials.2006.12.004.

Yang, Z., Duan, H., Mo, L., Qiao, H., & Li, X. (2010). The effect of the dosage of NT-3/chitosan carriers on the proliferation and differentiation of neural stem cells. *Biomaterials, 31*(18), 4846–4854. http://dx.doi.org/10.1016/j.biomaterials.2010.02.015.

Yang, Y., Gu, X., Tan, R., Hu, W., Wang, X., Zhang, P., et al. (2004). Fabrication and properties of a porous chitin/chitosan conduit for nerve regeneration. *Biotechnology Letters, 26*(23), 1793–1797. http://dx.doi.org/10.1007/s10529-004-4611-z.

Yang, F., Li, X., Cheng, M., Gong, Y., Zhao, N., Zhang, X., et al. (2002). Performance modification of chitosan membranes induced by gamma irradiation. *Journal of Biomaterials Applications, 16*(3), 215–226.

Yang, Z., Mo, L., Duan, H., & Li, X. (2010). Effects of chitosan/collagen substrates on the behavior of rat neural stem cells. *Science China. Life Sciences, 53*(2), 215–222. http://dx.doi.org/10.1007/s11427-010-0036-1.

Yeh, Y. S., Iriyama, Y., Matsuzawa, Y., Hanson, S. R., & Yasuda, H. (1988). Blood compatibility of surfaces modified by plasma polymerization. *Journal of Biomedical Materials Research, 22*(9), 795–818. http://dx.doi.org/10.1002/jbm.820220904.

Yoshii, S., Ito, S., Shima, M., Taniguchi, A., & Akagi, M. (2009). Functional restoration of rabbit spinal cord using collagen-filament scaffold. *Journal of Tissue Engineering and Regenerative Medicine, 3*(1), 19–25. http://dx.doi.org/10.1002/term.130.

Yoshii, S., Yamamuro, T., Ito, S., & Hayashi, M. (1987). In vivo guidance of regenerating nerve by laminin-coated filaments. *Experimental Neurology, 96*(2), 469–473.

Yu, L. M., Kazazian, K., & Shoichet, M. S. (2007). Peptide surface modification of methacrylamide chitosan for neural tissue engineering applications. *Journal of Biomedical Materials Research. Part A, 82*(1), 243–255. http://dx.doi.org/10.1002/jbm.a.31069.

Yu, L. M., Wosnick, J. H., & Shoichet, M. S. (2008). Miniaturized system of neurotrophin patterning for guided regeneration. *Journal of Neuroscience Methods, 171*(2), 253–263. http://dx.doi.org/10.1016/j.jneumeth.2008.03.023.

Yuan, Y., Zhang, P., Yang, Y., Wang, X., & Gu, X. (2004). The interaction of Schwann cells with chitosan membranes and fibers in vitro. *Biomaterials, 25*(18), 4273–4278. http://dx.doi.org/10.1016/j.biomaterials.2003.11.029.

Zahir, T., Nomura, H., Guo, X. D., Kim, H., Tator, C., Morshead, C., et al. (2008). Bioengineering neural stem/progenitor cell-coated tubes for spinal cord injury repair. *Cell Transplantation, 17*(3), 245–254.

Zhang, P. X., Jiang, B. G., Zhao, F. Q., Fu, Z. G., Zhang, D. Y., Du, C., et al. (2005). Chitin biological tube bridging the peripheral nerve with a small gap. *Zhonghua Wai Ke Za Zhi, 43*(20), 1344–1347.

Zhang, H., & Neau, S. H. (2001). In vitro degradation of chitosan by a commercial enzyme preparation: Effect of molecular weight and degree of deacetylation. *Biomaterials, 22*(12), 1653–1658.

Zheng, L., & Cui, H. F. (2012). Enhancement of nerve regeneration along a chitosan conduit combined with bone marrow mesenchymal stem cells. *Journal of Materials Science. Materials in Medicine, 23*(9), 2291–2302. http://dx.doi.org/10.1007/s10856-012-4694-3.

Zhu, Y., Gao, C., He, T., Liu, X., & Shen, J. (2003). Layer-by-layer assembly to modify poly(l-lactic acid) surface toward improving its cytocompatibility to human endothelial cells. *Biomacromolecules, 4*(2), 446–452. http://dx.doi.org/10.1021/bm025723k.

Zielinski, B. A., & Aebischer, P. (1994). Chitosan as a matrix for mammalian cell encapsulation. *Biomaterials, 15*(13), 1049–1056.

Zuo, Y. Y., Alolabi, H., Shafiei, A., Kang, N., Policova, Z., Cox, P. N., et al. (2006). Chitosan enhances the in vitro surface activity of dilute lung surfactant preparations and resists albumin-induced inactivation. *Pediatric Research, 60*(2), 125–130. http://dx.doi.org/10.1203/01.pdr.0000227558.14024.57.

CHAPTER TWO

Interfaces with the Peripheral Nerve for the Control of Neuroprostheses

Jaume del Valle, Xavier Navarro[1]
Department of Cell Biology, Physiology and Immunology, Faculty of Medicine, Institute of Neurosciences, Universitat Autònoma de Barcelona, Centro de Investigación Biomédica en Red sobre Enfermedades Neurodegenerativas (CIBERNED), Bellaterra, Spain
[1]Corresponding author: e-mail address: xavier.navarro@uab.cat

Contents

Abstract

Nervous system injuries lead to loss of control of sensory, motor, and autonomic functions of the affected areas of the body. Provided the high amount of people worldwide suffering from these injuries and the impact on their everyday life, numerous and different neuroprostheses and hybrid bionic systems have been developed to restore or partially mimic the lost functions. A key point for usable neuroprostheses is the electrode that interfaces the nervous system and translates not only motor orders into electrical outputs that activate the prosthesis but is also able to transform sensory information detected by the machine into signals that are transmitted to the central nervous system. Nerve electrodes have been classified with regard to their invasiveness in extraneural, intraneural, and regenerative. The more invasive is the implant the more selectivity of interfacing can be reached. However, boosting invasiveness and selectivity may also heighten nerve damage. This chapter provides a general overview of nerve electrodes as well as the state-of-the-art of their biomedical applications in neuroprosthetic systems.

International Review of Neurobiology, Volume 109
ISSN 0074-7742
http://dx.doi.org/10.1016/B978-0-12-420045-6.00002-X
63

1. GENERAL INTRODUCTION

Loss of function of denervated areas due to nerve injury or loss of a limb severely affects the ability to perform activities of daily living (ADL) and can significantly reduce quality of life of people who suffer from these conditions. Therefore, development of new approaches to partially mimic or restore the lost functions is an active field of research. If the connection between the central nervous system (CNS) and the target organ is no longer available but the muscles responsible for the lost action are still preserved, several systems with functional electrical stimulation (FES) have been developed and implanted to artificially stimulate the remaining organs or nerves, thus achieving the functions that could be carried out before the lesion. After limb loss due to amputation or congenital, the use of artificial prostheses is intended to substitute the missing parts of the body and recover its previous normal roles. The use of cosmetic prostheses can help to cope with the psychological distress produced by the missing body part improving patient's own image and self-assurance. However, with such prostheses, no restoration of sensorimotor lost functions is achieved. Thus, an ideal prosthesis would be that which (1) looks and feels familiar for the patient as happens with the aesthetic prostheses, (2) is thought controlled by the own patient, (3) allows for multiple degrees of freedom to resemble as much as possible the movements of the lost limb, and (4) provides the subject with sensory information from sensors located in the prosthesis. In order to be able to provide a natural-like control of the prosthesis, an interface is needed between the peripheral nervous system (PNS) and the machine. Thus, an electrode placed in the severed proximal nerves might selectively record motor efferent commands from the brain and transform these impulses into electrical signals that activate the motors of the prosthesis. Under the best circumstances, the electrode should also be able to translate the signals of the sensors in the device into electrical inputs and properly transform this information into sensory feedback (Micera & Navarro, 2009). Thus, nerve electrodes should bidirectionally and selectively interface different specific motor and sensory pathways with the best possible long-term biocompatibility. The main goal of this chapter is to review the current electrodes and strategies available for the development of advanced nerve–machine interface systems.

2. STRUCTURE AND FUNCTION OF THE PERIPHERAL NERVES

The neurons of the PNS are responsible for receiving and emitting impulses to connect through nerves of the CNS with the outside world (somatic PNS) or the own body (autonomic PNS). Peripheral nerves are formed by bundles of fascicles grouped by the epineurium, a layer of connective tissue that also contains blood vessels. Nerve fascicles are surrounded by another layer of connective tissue, the perineurium, and contain groups of axons that are embedded within a matrix of endoneurium. The fascicles are functionally and somatotopically organized to provide muscles, sensory organs, or viscera with motor information and to record sensory stimuli. Peripheral axons can be unmyelinated or myelinated, and the latter subdivided into different classes according to their diameter. The velocity of impulse conduction and the excitability for electrical stimulation are directly related to the size of the axons.

Somatic motor neurons have their soma located in the ventral horn of the spinal cord and are responsible for controlling the skeletal muscles of the body. Individual motor axons exit the spinal cord to innervate several muscle fibers constituting a motor unit. An action potential in a motoneuron travels through the axon jumping from node to node up to the axon terminals where, through the neuromuscular junction, it stimulates the muscle fibers of the motor unit producing their contraction. Thus, to increase muscle strength, several motor units need to be recruited, being the slow fatigue-resistant motor units, the first ones to be activated in a muscle contraction, whereas large fast-fatigue motor units are activated only when high levels of tension are needed. On the other hand, sensory neurons, whose soma are in the spinal ganglia, carry stimuli from different organs to the CNS. Changes in the environment of the skin, muscles, and joints that are detected by specific membrane receptors produce action potentials that travel along sensory myelinated or unmyelinated axons to bring mechanical, proprioceptive, thermal, or pain information to the CNS. Intensity of stimulation is coded by the frequency of action potentials and the number of sensory afferents recruited.

After peripheral nerve injury, the control of motor, sensory, and autonomic functions of the denervated areas can be lost. The regeneration of a severed nerve requires a complex and coordinated process in which axotomized neurons change to a regrowing mode, the distal stump debris are cleared by Wallerian degeneration, and the cut axons regenerate along the distal nerve stump. However, axon reconnection is far from optimal and target reinnervation is not selective at all after complete nerve transection (Allodi, Udina, & Navarro, 2012). Following injuries in proximal sites of the limbs, the slow rate of axonal regeneration implies chronic denervation and atrophy of target organs. Furthermore, if reconnection of proximal and distal stumps cannot be achieved, by direct suture or interposition of a graft, the cut axons are not able to elongate distally and aberrantly grow forming a proximal neuroma. The same situation is obvious after limb amputation. Moreover, regenerated or amputated nerves show an increase in excitability, causing undesired complications such as hyperreflexia, hyperalgesia, and pain.

3. TYPES OF NERVE ELECTRODES: AN OVERVIEW

Different types of electrodes have been developed to interface the PNS in order to record electrical activity from and/or to stimulate the nerve fibers for different biomedical applications (Navarro et al., 2005; Schultz & Kuiken, 2011). Most interfaces are implanted around or within a peripheral nerve or spinal root in order to have low tissue resistance, therefore reducing the intensity needed for stimulation and enhancing the signal-to-noise ratio for recordings. Nerve electrodes can be classified into three main classes depending on nerve invasiveness: extraneural, intraneural, and regenerative (Fig. 2.1). With increasing invasiveness of the implant, higher selectivity of stimulation of individual nerve fibers may be reached and lower intensity is needed as the distance from the electrode to individual axons is reduced; likewise, reducing this distance has a positive influence on the quality of the recorded signals. However, selectivity comes with a price as the more invasive is the electrode, the more potential damage to the nerve is prone to be done. For example, extraneural electrodes, such as cuff and epineurial ones, provide simultaneous interface with many axons in the nerve resulting in poor selectivity but with very little nerve damage, whereas intraneural electrodes inserted in the nerve may interface discrete groups of axons within a fascicle reaching very good selectivity but with higher risk of nerve damage.

3.1. Extraneural electrodes

Cuff electrodes (Fig. 2.1A) are made of a cylindrical sheath that is wrapped longitudinally around the nerve with two or more electrode sites in the

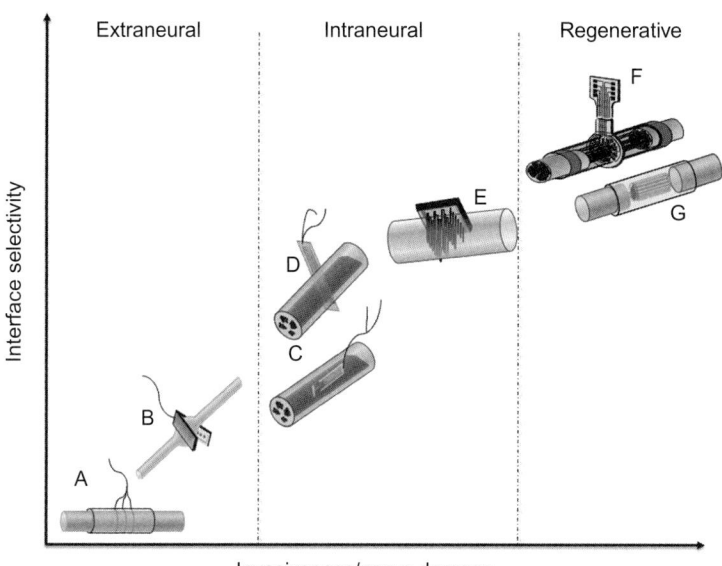

Figure 2.1 Electrodes used to interface peripheral nerves classified according to their invasiveness and selectivity. Images show examples of (A) cuff electrode, (B) flat interface nerve electrode (FINE), (C) longitudinal intrafascicular electrode (LIFE), (D) transverse intrafascicular multichannel electrode (TIME), (E) multielectrode array (USEA), (F) sieve electrode, and (G) microchannel electrode. (For color version of this figure, the reader is referred to the online version of this chapter.)

lumen of the tube (Hoffer & Loeb, 1980). Cuff electrodes allow for precise positioning and significantly reduce stimulus intensity compared to surface and epimysial electrodes (Loeb & Peck, 1996), since the cuff-insulating sheath limits current leak out of the cuff-nerve space. Compared to intraneural electrodes, the surrounding approach of cuff electrodes reduces their selectivity, making them able to record and stimulate only sensorimotor large myelinated fibers and predominantly those located at superficial locations (Badia, Boretius, Andreu, et al., 2011). On the other hand, the reduced invasiveness of these electrodes makes them easier to handle and safer to implant (Naples, Mortimer, & Yuen, 1990). In order to improve cuff's selectivity and performance, several strategies, such as multisite cuff electrodes (Navarro, Valderrama, Stieglitz, & Schüttler, 2001; Tarler & Mortimer, 2004; Veraart, Grill, & Mortimer, 1993; Walter et al., 1997), innovative cuff structures (Tyler & Durand, 1997), complex modes of stimulation (Grill & Mortimer, 1996; Navarro et al., 2001), and advanced processing algorithms (Raspopovic, Carpaneto, Udina, Navarro, & Micera, 2010;

Tesfayesus & Durand, 2006), have been developed. The reduced size and thickness of thin-film polymer cuffs make feasible the implantation of several small cuffs around different fascicles or branches of a peripheral nerve, consequently achieving selective stimulation of a higher number of targets (Stieglitz, 2007). Cuff electrodes have a long record of use in FES systems for human applications, such as in devices to correct foot-drop, allow hand grasping, and control of micturition, with successful outcomes over years of use (Brindley, 1994; Lyons, Sinkjær, Burridge, & Wilcox, 2002; Waters, McNeal, Faloon, & Clifford, 1985).

Flat interface nerve electrode (FINE, Fig. 2.1B) is a design variation of the cuff electrode developed by Durand and coworkers (Leventhal & Durand, 2003; Tyler & Durand, 2002). The FINE applies a small pressure to the nerve to enlarge its cross-section and thus increases the nerve surface area. Such flattening of the nerve displaces the axons from the center to the surface and closer to the electrode active sites. Studies in laboratory animals with FINEs implanted over 1–3 months showed that electrodes applying moderate and small forces did not cause detectable nerve damage, but high reshaping forces induced nerve lesion (Tyler & Durand, 2003). Studies in humans have revealed that FINEs placed on the femoral nerve trunk can selectively and independently activate the different muscles innervated by the femoral nerve (Schiefer, Polasek, Triolo, Pinault, & Tyler, 2010).

3.2. Intraneural electrodes

Intraneural electrodes are implanted within nerve fascicles. They show a better selectivity than extraneural electrodes as they may have closer contact with different fascicles in both superficial and deep-nerve locations. Therefore, the threshold to stimulate the axons is reduced, and both the selectivity and the signal-to-noise are improved (Badia, Boretius, Andreu, et al., 2011). On the other hand, their invasiveness is higher than the extraneural electrodes as the implantation itself may cause more damage to the nerve. An important issue limiting the usability of intraneural electrodes is the long-term stability of the contact between the probe and the nerve fibers. Shape memory alloys applied as smart actuators in the electrode and development of biological methods to reduce the fibrotic reaction around the electrode are promising methods to improve the performance for long-term application (Bossi et al., 2007).

Longitudinal intrafascicular electrodes (LIFEs) (Fig. 2.1C) are constructed from thin, insulated conducting wires (such as Pt–Ir or metalized polymers) or polymer filaments that are inserted longitudinally into individual nerve fascicles, so as to lay in-between and parallel to the nerve fibers (Lawrence, Dhillon, Jensen, Yoshida, & Horch, 2004; Yoshida & Horch, 1993; Yoshida & Stein, 1999). The intrafascicular location makes them able to record and/or stimulate only small groups of axons providing high selectivity. An evolution of these electrodes is the thin-film LIFE (tfLIFE) based on a thin micropatterned polyimide substrate. This highly flexible substrate filament is folded in half, and each side can host a number of active sites within a small surface, thus allowing selective multiunit nerve recording and stimulation (Navarro et al., 2007). Although initially developed for use with FES, LIFE electrodes are being used also for bidirectional interfacing peripheral nerves after lesions and amputations. After 3 months, implantation in the rat sciatic nerve tfLIFE did not induce noticeable abnormalities in nerve function and morphology, and only a mild inflammatory reaction around the electrode was observed (Lago, Udina, Ramachandran, & Navarro, 2007). Several cases have been reported of LIFEs implanted in human amputees, providing evidence of their effective use for controlling advanced hand prostheses, and for delivering sensory feedback to the subject (Dhillon, Krüger, Sandhu, & Horch, 2005; Rossini et al., 2010).

Transverse intrafascicular multichannel electrodes (TIMEs) (Fig. 2.1D) are designed to be implanted transversally into the nerve in order to access different groups of nerve fibers. In comparison with LIFE that has high selectivity to interface a small population of nerve fibers within one fascicle, the TIME is able to record or stimulate different subsets of axons in various fascicles over the nerve cross-section and obtain a reasonable spatial selectivity (Boretius et al., 2010). Studies in rats have shown that after implantation in the sciatic nerve, it was possible to selectively activate different muscles innervated by distinct branches of the nerve or even within the same fascicle, while LIFE was only able to stimulate one of each fascicle (Badia, Boretius, Andreu, et al., 2011). In addition, no significant axonal damage was found 2 months after implantation in the sciatic nerve, suggesting the TIME as a good candidate for chronic implantation even in small peripheral nerves (Badia, Boretius, Pascual-Font, et al., 2011).

The *self-opening intrafascicular neural interface* (SELINE) is an evolution of tfLIFE and TIME electrodes. The electrode is composed of a looped polyimide thin film body and lateral wings with active sites on them (Cutrone et al., 2011). After insertion of the structure in the nerve, it is gently pulled away allowing the wings to open transversely. Therefore, the wings move within fascicles, offering a second dimension for contacting more axons from different fascicles. The reduced size of the wings would make the SELINE less prone to produce fibrotic encapsulation and electrical insulation.

Multielectrode arrays (MEAs) (Fig. 2.1E) are electrodes composed of a base made of silicon, ceramic, polyimide, or glass, and tens of needles with electrode tips made of carbon, gold, platinum, or iridium oxide (Nordhausen, Maynard, & Normann, 1996; Zhang, Peng, Qi, Gao, & Zhang, 2009). MEAs are inserted transversely into the nervous system thus providing multisite recording and stimulation. On the other hand, the high number of electrical contacts may generate some neural damage produced not only by the rigid structure of the electrodes in contact with the nervous tissue but also by tethering forces during movements by lead wires. Over the past decade, these electrodes have been mainly used as CNS microinterfaces to selectively record or stimulate the brain cortex, allowing communication of paralyzed patients to control computers and robotic assistive devices during years after implantation (Hochberg et al., 2012, 2006; Velliste, Perel, Spalding, Whitford, & Schwartz, 2008). However, MEAs have also been used in peripheral nerves in experimental works with cats (Branner & Normann, 2000) and also in human volunteers (Warwick et al., 2003). A modified version of the MEA (named Utah Slanted Electrode Array, USEA) has been developed. The array is composed of electrodes of different length in order to reduce the number of redundant electrodes. Therefore, access to more fascicles within the nerve is provided, allowing low-current highly selective stimulation of motor fibers (Branner, Stein, Fernandez, Aoyagi, & Normann, 2004; Branner, Stein, & Normann, 2001) and selective recording of single-unit responses (Clark, Ledbetter, Warren, & Harrison, 2011) during several months after nerve implantation. Nevertheless, chronic studies reported poor stability of recorded signals over time (Branner et al., 2004; Warwick et al., 2003). Promising results have been obtained both *in vitro* and *in vivo* with newly developed flexible and stretchable MEAs. These electrodes are polyimide- or silicone based and they can deform in 3D along with the nerve, thus diminishing traumatic injuries produced by mechanical tension in the nerve interface (Lacour et al., 2010).

3.3. Regenerative electrodes

Regenerative electrodes represent a different approach to interface the PNS and are probably the most invasive electrodes but also the ones that might offer the highest level of selectivity. These electrodes are not thought to be implanted in an intact nerve but to have the nerve grow through them. The first and most investigated design of a regenerative electrode was the sieve electrode (Fig. 2.1F), composed of an array of via holes with electrodes built around them and placed in the gap between stumps of a transected nerve. Thus, the axons regenerate through the holes, and the electrodes can interface the small bunch of axons of every single hole reaching a high level of selectivity.

The most ambitious utilization of sieve electrodes is the implantation in the sectioned nerves of an amputee so as to interface the axons that formerly innervated the severed limb for a bidirectional interface with the prosthetic limb. However, it cannot be discarded the even more challenging use of controlling signal rerouted from the regenerated nerve to the adequate distal target after simple nerve transection (Rosen, Grosser, & Hentz, 1990). Sieve electrodes have not been implanted in humans yet, but several studies have verified nerve regeneration through these electrodes with successful interfacing of the regenerated fibers in frogs (Kovacs et al., 1994), fish (Mensinger et al., 2000), rats (Navarro et al., 1996, 1998), rabbits (Kawada et al., 2004), and cats (Panetsos, Avendaño, Negredo, Castro, & Bonacasa, 2008). Regenerative electrodes not only grant selective stimulation of small groups of regenerated fibers (Lago, Udina, et al., 2007) but also allow providing receptive signals from different sensory areas to the interface (Navarro et al., 1998). Despite full nerve regeneration is not achieved due to the obstacle of the sieve for growing axons, sieve electrodes have shown promising results after 30 months implant in cats, as only minor electrical changes were found in regenerated nerves and sound levels of functionality were reached (Panetsos et al., 2008). However, extensive research done with these electrodes has raised some questions that still need to be solved. Sieve electrodes are thought to make the axons grow through small holes limiting the possibility to use these electrodes in acute experiments (Navarro et al., 2005). Long-term studies have shown signs of axonopathy in some regenerated fibers due to compression after 6 months in rats (Lago, Ceballos, Rodríguez, Stieglitz, & Navarro, 2005). In addition, axons do not grow following the same fascicle topography than in the intact nerve and this disorganization can difficult optimal interfacing of distinct fascicles. Moreover, small unmyelinated axons have a higher growing capacity in comparison with larger myelinated axons through the sieves; this favors the smaller axons to grow through the center of the electrode submitting larger axons to the periphery or even outside the interface thus preventing their possibility to be properly contacted (Castro, Negredo, & Avendaño, 2008; Lago, Udina, et al., 2007).

Several alternative designs have been proposed for improving the amount of axonal regeneration through nonobstructive, regenerative multielectrodes. A simple design was to insert 18 needle electrodes transversally in a nerve guide used to bridge the sectioned nerve. Initial studies *in vivo* showed that such electrode arrays placed in the path of regenerating nerve fibers allowed the recording of action potentials from as early as 8 days postimplantation to as long as 3 months in a low proportion of the animals (Garde, Keefer, Botterman, Galvan, & Romero, 2009).

The natural tendency of regenerating axons to form new small fascicles makes optimal selective recording or stimulation difficult to accomplish. Moreover, at difference with the normal topography of the peripheral nerve, regenerated fascicles contain a mix of sensory and motor fibers directed to diverse targets, with loss of the normal somatotopic map. *In vitro* attempts have been made to guide regenerating axons into small fascicles by promoting neurite growth into microchannels with stepping bifurcations (Wieringa, Wiertz, De Weerd, & Rutten, 2010), and to separate distinct functional types of regenerating axons by exposure to different neurotrophic factors in Y-shape channels (Lotfi, Garde, Chouhan, Bengali, & Romero-Ortega, 2011).

New alternatives in the development of regenerative electrodes have explored the potential benefits of designing, facilitating regenerative scaffolds (Clements et al., 2013) and the so-called microchannel electrodes (Fig. 2.1G; FitzGerald, Lacour, & Fawcett, 2008). These electrodes could be thought as an evolution of sieve electrodes in which, instead of growing through holes, the axons grow via thin, narrow parallel tubes with embedded electrodes. In order to resemble the topographical organization of a healthy nerve, the tubes can incorporate guidance cues to improve nerve regeneration directing axonal growth thus mimicking the fascicle distribution of a peripheral nerve (Lundborg & Kanje, 1996). The long contacts the electrodes make with the axon along the microchannel facilitate that every electrode will eventually contact one or more nodes of Ranvier, increasing the amplitude of recorded action potentials (Lacour et al., 2009) and allowing for stimulation of single motor units with lower thresholds than with other types of electrodes (FitzGerald et al., 2012). Some studies in rats reported that these new approaches allow a high selectivity in recording and stimulation of the regenerated axons (Clements et al., 2013; Delivopoulos, Chew, Minev, Fawcett, & Lacour, 2012), although successful regeneration occurred only in a low proportion of the animals (FitzGerald et al., 2012).

4. BIOMEDICAL APPLICATIONS OF NERVE INTERFACES

Both PNS and CNS dysfunctions by disease, trauma, or injury produce impairments in the ADL. Applications of nerve interfaces linked with neuroprosthetic systems can adopt two main approaches: (1) replacement of central control after CSN injuries by FES systems that stimulate the intact peripheral nerves or muscles, generating movements or functions that reproduce normal actions and (2) replacement of injured peripheral nerves by connecting proximal nerve segments with denervated muscles or artificial prostheses substituting lost parts of the body (Fig. 2.2). Regain of control in micturition and defecation, walking improvement after gait impairment or paraplegia, and also prosthetic substitution after loss of a limb are some of the topics that have been addressed by trying to electronically interface remaining nerve pathways and using different prosthetic devices. Moreover, not only efferent signals are targeted to elicit functional activities, but also recording of afferent signals is aimed to provide sensory information and produce a bidirectional interface with the nervous system, and electrical modulation of disturbed neural circuits is being increasingly applied to different neural disorders in exciting and challenging biomedical applications.

4.1. Applications for CNS-injured patients

Both brain and spinal cord injuries induce significant neurological and functional disability from which full recovery is far from being expected. FES relies on electrical stimulation of a nerve or a muscle that either has lost

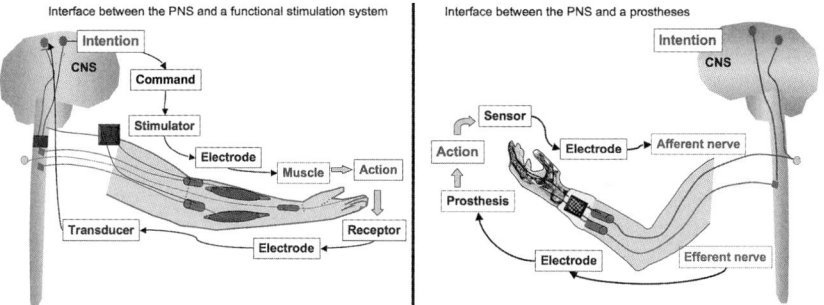

Figure 2.2 Schematic representation of the two main application modes of peripheral nerve interface systems, represented for the control of the hand by a functional electrical stimulation system in a tetraplegic case, and for the control of a bionic prosthesis in an amputee. (For color version of this figure, the reader is referred to the online version of this chapter.)

the connection with the brain or the spinal cord or that no longer is receiving appropriate signals for its normal function. FES of target organs aims to support or replace the lost central control of CNS-injured patients and helps them to improve quality of life and independency. With the help of closed-loop control, FES systems can process different outputs to adapt the stimulation to a better performance (Micera & Navarro, 2009). Different FES systems have been successfully developed and implanted in patients to enhance various aspects of everyday activities.

Sacral root stimulation. Implantable sacral nerve stimulation is a minimally invasive and durable procedure for patients with urinary and fecal incontinence or neurogenic overactivity who are refractory to conservative therapy. It involves stimulation of the sacral nerves with electrodes, cuff, or book type, implanted around sacral spinal roots and connected to a pulse generator. Additional neuromodulation via stimulation of the afferent nerves helps to inhibit inappropriate neural reflex behavior. This technique has been shown to improve incontinence, retention, and reduce urinary infections (Brindley, 1994; Creasey & Craggs, 2012; Mayer & Howard, 2008). The sacral anterior root stimulation has also been reported to reduce constipation, improve bowel function, and satisfaction in spinal cord-injured patients (Vallès, Rodríguez, Borau, & Mearin, 2009). Regarding sexual dysfunction, stimulation of parasympathetic efferents is able to produce penile erection and seminal emission, whereas reflex erection and ejaculation can also be produced by stimulation of afferent nerves (Creasey & Craggs, 2012).

Ventilatory pacing. Diaphragm pacing by electrical stimulation of the phrenic nerve can provide significant benefits to selected patients with respiratory paralysis, such as those with high cervical tetraplegia. The phrenic nerve, stimulated by epineural or cuff electrodes, produces diaphragm-paced contractions for ventilatory assistance (Creasey et al., 1996). Multichannel electrodes have the advantage of stimulating different portions of the nerve each time with the intention of reducing muscle fatigue (Jarosz, Littlepage, Creasey, & McKenna, 2012).

Correction of foot drop. Lower limb neuroprostheses in the form of peroneal nerve stimulators are effective in improving the gait of stroke patients with foot-drop. Electrical stimulation of the peroneal nerve during the advance phase of the affected leg produces contraction of foot dorsiflexor muscles lifting the foot from the ground, thereby improving the walking ability (Lyons et al., 2002). Feedback to improve control of the stimulation has been achieved by using mechanoreceptive signals from the sole recorded with a cuff electrode implanted in the sural nerve at the ankle (Haugland & Sinkjaer, 1999), or force sensors measuring the components of moment generated at the ankle joint (Kottink et al., 2004).

Walking assistance prostheses. Several FES systems have been designed and implanted for assisting standing and walking in paraplegic and stroke patients. Bilateral programmed systems excite the quadriceps muscle or the femoral nerve to maintain knee extension for stance in one lower limb, whereas the other stimulates the common peroneal nerve at a high strength to elicit a flexion reflex that produces flexion of the hip and knee for the swing phase in the other limb (Stein & Mushahwar, 2005). Some systems use surface electrodes (Graupe, Cerrel-Bazo, Kern, & Carraro, 2008; Popovic & Keller, 2005), but more advanced ones take use of a variety of epimysial, intramuscular, and nerve electrodes implanted in the appropriate muscles or nerves (Bailey et al., 2010; Guiraud, Stieglitz, Koch, Divoux, & Rabischong, 2006). Systems with more electrodes allow for more normal pattern of walking and may reduce the excessive metabolic demand and muscle fatigue; moreover, neural stimulation proved to be more efficient, require less energy, and provide more selective stimulation than muscular stimulation. Electromyography (EMG) signals recorded from partially controlled muscles can be used to trigger FES-assisted gait initiation that results more coordinated and stable than with switch-triggered systems (Dutta, Kobetic, & Triolo, 2009).

Hand grasping prostheses. Restoration of hand function in tetraplegic and stroke patients is achieved by upper extremity neuroprostheses that use FES to power hand and arm muscles. A variety of devices send signals via a small external controller and transmitting coil to an implanted stimulator. The stimulator activates selected arm and hand muscles via implanted wires and muscular electrodes. Palmar and lateral grasp, among other functions, can be reliably restored, leading to significant improvements in ADL (Peckham et al., 2001; Popovic, Curt, Keller, & Dietz, 2001; Rupp & Gerner, 2004). The hand closure and opening may be commanded using a position sensor placed on the shoulder of the subject, a push button or pressure sensors. Recent advances have demonstrated improved outcomes of hand grasping by incorporating start control and proportional control of grasp strength from EMG signals recorded in shoulder and arm muscles (Kilgore et al., 2008), or by providing sensory feedback to the system by neural signals recorded with cuff electrodes around digital nerves (Inmann & Haugland, 2004). Exciting ongoing research aims to bring such a hand neuroprosthesis under volitional control from either noninvasive

EEG signals (Millán et al., 2010; Müller-Putz, Scherer, Pfurtscheller, & Rupp, 2005) or from recordings with microelectrodes permanently implanted in the motor cortex (Ethier, Oby, Bauman, & Miller, 2012).

In addition to the assistance in the direct actions intended, FES systems are also valuable to raise the general well-being of injured patients, reduce morbidity derived from paralysis, and improve their self-esteem (Ragnarsson, 2007).

4.2. Applications for the control of prostheses

Although hand transplantation has been successfully achieved, this option still seems restricted to a selected group of patients (Schuind, Abramowicz, & Schneeberger, 2007), leaving the vast majority of people who have lost a limb with either functional or cosmetic prostheses or even with no intervention. Hence, although several high-tech prosthetic devices have been developed with up to 22 degrees of freedom (Cipriani, Controzzi, & Carrozza, 2011; Weir et al., 2008), there is still the need of an available and satisfactory cybernetic prosthesis that would provide enough operability and appropriate sensory feedback to partially restore the functionality of the lost hand.

The most common approach to control a mechanical hand is by placing surface electrodes onto intact muscles in the proximal limb stump. When these muscles are voluntary activated by the amputee, the detected EMG signals are translated to perform predefined movements (Scott & Parker, 1988). This approach has been limited to the control of two degrees of freedom, or movements, in the prosthesis. Progress in myoelectric control has increased by developments in the means of extracting information from the EMG signals. Different computational methods for pattern recognition of myoelectric signals from uninjured or related musculature of the lost limb are recognized and patterned to externally control multigrasping tasks by the prosthesis (Fougner, Stavdahl, Kyberd, Losier, & Parker, 2012; Micera et al., 2011; Park & Lee, 1998).

In recent years, research into the control of upper limb prostheses has attempted to go far beyond conventional prosthetic treatment by the use of novel interfaces to the nervous system (Micera et al., 2008; Schultz & Kuiken, 2011). An interesting approach to command a multijoint artificial limb is targeted reinnervation (Kuiken, Marasco, Lock, Harden, & Dewald, 2007; Kuiken et al., 2009). This procedure remains on the basis that the remaining nerves of an amputated limb are still viable and receive CNS signals. The amputated nerves are surgically reconnected to residual muscles that they reinnervate. Therefore, patients can send volitional orders to move the phantom limb via the nerves that now are stimulating the residual muscles, whose EMG activity is recorded by a surface electrode array and redirected to control artificial prostheses with several degrees of freedom, and that can also be processed by pattern recognition (Kuiken et al., 2009). Moreover, these amputees were even able to perceive tactile sensations of the missing limb when touched on the reinnervated skin areas (Sensinger, Schultz, & Kuiken, 2009).

Further advances under testing rely on directly interfacing the severed nerves that subserved the amputated limb with multipolar electrodes that may record motor commands and stimulate sensory feedback, thus reestablishing a bidirectional link between the nervous system and the prosthesis in a physiological manner. Recent investigations have reported that a cybernetic prosthesis can be commanded by interfacing peripheral nerves with intraneural electrodes. Several LIFEs were implanted in the median and ulnar nerves of chronic amputees, allowing to perform three distinct hand movements under voluntary control, and providing sensory feedback from the hand prosthesis that allowed perceptions of joint position and object recognition (Dhillon, Lawrence, Hutchinson, & Horch, 2004; Dhillon et al., 2005; Horch, Meek, Taylor, & Hutchinson, 2011; Micera et al., 2011; Rossini et al., 2010). Still limited but promising experiences support the view that intraneural interfaces may also allow a refined link with external teleoperated robotic devices (Warwick et al., 2003).

5. CONCLUSION

The possibility of interfacing and controlling artificial machines with biological signals has a long history of multidisciplinary research. The rapid growth of knowledge and applications in the fields of neural regeneration and neuroprosthesis has prompted to identify new areas for converging research, which are evolving in recent years. They include, among others, enhancement of axonal regeneration and injury healing by electrical modulation, development of hybrid neural interfaces combining artificial and biological elements, investigation of patterned neural activity in regulating dysfunctional neuronal circuits, and promoting neural plasticity for improving rehabilitation (Andrews, 2009; Aravamudhan & Bellamkonda, 2011; Grill et al., 2001).

REFERENCES

Allodi, I., Udina, E., & Navarro, X. (2012). Specificity of peripheral nerve regeneration: Interactions at the axon level. *Progress in Neurobiology, 98*, 16–37.

Andrews, R. J. (2009). Neuromodulation: Deep brain stimulation, sensory neuroprostheses, and the neural–electrical interface. *Progress in Brain Research, 180*, 127–139.

Aravamudhan, S., & Bellamkonda, R. V. (2011). Toward a convergence of regenerative medicine, rehabilitation, and neuroprosthetics. *Journal of Neurotrauma, 28*, 2329–2347.

Badia, J., Boretius, T., Andreu, D., Azevedo-Coste, C., Stieglitz, T., & Navarro, X. (2011). Comparative analysis of transverse intrafascicular multichannel, longitudinal intrafascicular and multipolar cuff electrodes for the selective stimulation of nerve fascicles. *Journal of Neural Engineering, 8*, 036023.

Badia, J., Boretius, T., Pascual-Font, A., Udina, E., Stieglitz, T., & Navarro, X. (2011). Biocompatibility of chronically implanted transverse intrafascicular multichannel electrode (TIME) in the rat sciatic nerve. *IEEE Transactions on Biomedical Engineering, 58*, 2324–2332.

Bailey, S. N., Hardin, E. C., Kobetic, R., Boggs, L. M., Pinault, G., & Triolo, R. J. (2010). Neurotherapeutic and neuroprosthetic effects of implanted functional electrical stimulation for ambulation after incomplete spinal cord injury. *Journal of Rehabilitation Research and Development, 47*, 7–16.

Boretius, T., Badia, J., Pascual-Font, A., Schuettler, M., Navarro, X., Yoshida, K., et al. (2010). A transverse intrafascicular multichannel electrode (TIME) to interface with the peripheral nerve. *Biosensors and Bioelectronics, 26*, 62–69.

Bossi, S., Menciassi, A., Koch, K. P., Hoffmann, K. P., Yoshida, K., Dario, P., et al. (2007). Shape memory alloy microactuation of tf-lIFes: Preliminary results. *IEEE Transactions on Biomedical Engineering, 54*, 1115–1120.

Branner, A., & Normann, R. A. (2000). A multielectrode array for intrafascicular recording and stimulation in sciatic nerve of cats. *Brain Research Bulletin, 51*, 293–306.

Branner, A., Stein, R. B., Fernandez, E., Aoyagi, Y., & Normann, R. A. (2004). Long-term stimulation and recording with a penetrating microelectrode array in cat sciatic nerve. *IEEE Transactions on Biomedical Engineering, 51*, 146–157.

Branner, A., Stein, R. B., & Normann, R. A. (2001). Selective stimulation of cat sciatic nerve using an array of varying-length microelectrodes. *Journal of Neurophysiology, 85*, 1585–1594.

Brindley, G. S. (1994). The first 500 patients with sacral anterior root stimulator implants: General description. *Spinal Cord, 32*, 795–805.

Castro, J., Negredo, P., & Avendaño, C. (2008). Fiber composition of the rat sciatic nerve and its modification during regeneration through a sieve electrode. *Brain Research, 1190*, 65–77.

Cipriani, C., Controzzi, M., & Carrozza, M. C. (2011). The SmartHand transradial prosthesis. *Journal of Neuroengineering and Rehabilitation, 8*, 29.

Clark, G. A., Ledbetter, N. M., Warren, D. J., & Harrison, R. R. (2011). Recording sensory and motor information from peripheral nerves with Utah Slanted Electrode Arrays. In *2011 Annual International Conference of the IEEE Engineering in Medicine and Biology Society* (pp. 4641–4644).

Clements, I. P., Mukhatyar, V., Srinivasan, A., Bentley, J. T., Andreasen, D. S., & Bellamkonda, R. V. (2013). Regenerative scaffold electrodes for peripheral nerve interfacing. *IEEE Transactions on Neural System Rehabilitation Engineering, 21*, 554–566.

Creasey, G. H., & Craggs, M. D. (2012). Functional electrical stimulation for bladder, bowel, and sexual function. *Spinal Cord Injuries E-Book: Handbook of Clinical Neurology Series, 109*, 247.

Creasey, G., Elefteriades, J., DiMarco, A., Talonen, P., Bijak, M., Girsch, W., et al. (1996). Electrical stimulation to restore respiration. *Journal of Rehabilitation Research and Development*, *33*, 123–132.

Cutrone, A., Sergi, P. N., Bossi, S., & Micera, S. (2011). Modelization of a self-opening peripheral neural interface: A feasibility study. *Medical Engineering & Physics*, *33*, 1254–1261.

Delivopoulos, E., Chew, D. J., Minev, I. R., Fawcett, J., & Lacour, S. (2012). Concurrent recordings of bladder afferents from multiple nerves using a microfabricated PDMS microchannel electrode array. *Lab on a Chip*, *12*, 2540–2551.

Dhillon, G. S., Krüger, T. B., Sandhu, J. S., & Horch, K. W. (2005). Effects of short-term training on sensory and motor function in severed nerves of long-term human amputees. *Journal of Neurophysiology*, *93*, 2625–2633.

Dhillon, G. S., Lawrence, S. M., Hutchinson, D. T., & Horch, K. W. (2004). Residual function in peripheral nerve stumps of amputees: Implications for neural control of artificial limbs. *Journal of Hand Surgery*, *29*, 605–615.

Dutta, A., Kobetic, R., & Triolo, R. J. (2009). Gait initiation with electromyographically triggered electrical stimulation in people with partial paralysis. *Journal of Biomechanical Engineering*, *131*, 81002.

Ethier, C., Oby, E. R., Bauman, M. J., & Miller, L. E. (2012). Restoration of grasp following paralysis through brain-controlled stimulation of muscles. *Nature*, *485*, 368–371.

FitzGerald, J. J., Lacour, S. P., & Fawcett, J. W. (2008). Recording with microchannel electrodes in a noisy environment. In *30th Annual International Conference of the IEEE Engineering in Medicine and Biology Society* (pp. 34–37).

FitzGerald, J. J., Lago, N., Benmerah, S., Serra, J., Watling, C. P., Cameron, R. E., et al. (2012). A regenerative microchannel neural interface for recording from and stimulating peripheral axons in vivo. *Journal of Neural Engineering*, *9*, 016010.

Fougner, A., Stavdahl, O., Kyberd, P. J., Losier, Y. G., & Parker, P. A. (2012). Control of upper limb prostheses: Terminology and proportional myoelectric control—A review. *IEEE Transactions on Neural Systems and Rehabilitation Engineering*, *20*, 663–677.

Garde, K., Keefer, E., Botterman, B., Galvan, P., & Romero, M. I. (2009). Early interfaced neural activity from chronic amputated nerves. *Frontiers in Neuroengineering*, *2*, 5.

Graupe, D., Cerrel-Bazo, H., Kern, H., & Carraro, U. (2008). Walking performance, medical outcomes and patient training in FES of innervated muscles for ambulation by thoracic-level complete paraplegics. *Neurological Research*, *30*, 123–130.

Grill, W. M., McDonald, J. W., Peckham, P. H., Heetderks, W., Kocsis, J., & Weinrich, M. (2001). At the interface: Convergence of neural regeneration and neural prostheses for restoration of function. *Journal of Rehabilitation Research and Development*, *38*, 633–640.

Grill, W. M., Jr., & Mortimer, J. T. (1996). The effect of stimulus pulse duration on selectivity of neural stimulation. *IEEE Transactions on Biomedical Engineering*, *43*, 161–166.

Guiraud, D., Stieglitz, T., Koch, K. P., Divoux, J. L., & Rabischong, P. (2006). An implantable neuroprosthesis for standing and walking in paraplegia: 5-year patient follow-up. *Journal of Neural Engineering*, *3*, 268.

Haugland, M., & Sinkjaer, T. (1999). Interfacing the body's own sensing receptors into neural prosthesis devices. *Technology for Health Care*, *7*, 393–399.

Hochberg, L. R., Bacher, D., Jarosiewicz, B., Masse, N. Y., Simeral, J. D., Vogel, J., et al. (2012). Reach and grasp by people with tetraplegia using a neurally controlled robotic arm. *Nature*, *485*, 372–375.

Hochberg, L., Serruya, S., Friehs, G., Mukand, J., Saleh, M., Caplan, A., et al. (2006). Neuronal ensemble control of prosthetic devices by a human with tetraplegia. *Nature*, *442*, 164–171.

Hoffer, J. A., & Loeb, G. E. (1980). Implantable electrical and mechanical interfaces with nerve and muscle. *Annals of Biomedical Engineering, 8*(4), 351–360.

Horch, K., Meek, S., Taylor, T. G., & Hutchinson, D. T. (2011). Object discrimination with an artificial hand using electrical stimulation of peripheral tactile and proprioceptive pathways with intrafascicular electrodes. *IEEE Transactions on Neural Systems and Rehabilitation Engineering, 19*, 483–489.

Inmann, A., & Haugland, M. (2004). Implementation of natural sensory feedback in a portable control system for a hand grasp neuroprosthesis. *Medical Engineering & Physics, 26*, 449–458.

Jarosz, R., Littlepage, M. M., Creasey, G., & McKenna, S. L. (2012). Functional electrical stimulation in spinal cord injury respiratory care. *Topics in Spinal Cord Injury Rehabilitation, 18*, 315–321.

Kawada, T., Zheng, C., Tanabe, S., Uemura, T., Sunagawa, K., & Sugimachi, M. (2004). A sieve electrode as a potential autonomic neural interface for bionic medicine. In *26th Annual International Conference of the IEEE Engineering in Medicine and Biology Society*, Vol. 2, (pp. 4318–4321).

Kilgore, K. L., Hoyen, H. A., Bryden, A. M., Hart, R. L., Keith, M. W., & Peckham, P. H. (2008). An implanted upper-extremity neuroprosthesis using myoelectric control. *Journal of Hand Surgery, 33*, 539–550.

Kottink, A. I., Buschman, H. P., Kenney, L. P., Veltink, P. H., Slycke, P., Bultstra, G., et al. (2004). The sensitivity and selectivity of an implantable two-channel peroneal nerve stimulator system for restoration of dropped foot. *Neuromodulation, 7*, 277–283.

Kovacs, G. T., Storment, C. W., Halks-Miller, M., Belczynski, C. R., Jr., Santina, C. D., Lewis, E. R., et al. (1994). Silicon-substrate microelectrode arrays for parallel recording of neural activity in peripheral and cranial nerves. *IEEE Transactions on Biomedical Engineering, 41*, 567–577.

Kuiken, T. A., Li, G., Lock, B. A., Lipschutz, R. D., Miller, L. A., Stubblefield, K. A., et al. (2009). Targeted muscle reinnervation for real-time myoelectric control of multifunction artificial arms. *Journal of the American Medical Association, 301*, 619–628.

Kuiken, T. A., Marasco, P. D., Lock, B. A., Harden, R. N., & Dewald, J. P. (2007). Redirection of cutaneous sensation from the hand to the chest skin of human amputees with targeted reinnervation. *Proceedings of the National Academy of Sciences, 104*, 20061–20066.

Lacour, S. P., Benmerah, S., Tarte, E., FitzGerald, J., Serra, J., McMahon, S., et al. (2010). Flexible and stretchable micro-electrodes for in vitro and in vivo neural interfaces. *Medical and Biological Engineering and Computing, 48*, 945–954.

Lacour, S. P., Fitzgerald, J. J., Lago, N., Tarte, E., McMahon, S., & Fawcett, J. (2009). Long micro-channel electrode arrays: A novel type of regenerative peripheral nerve interface. *IEEE Transactions on Neural Systems and Rehabilitation Engineering, 17*, 454–460.

Lago, N., Ceballos, D., Rodríguez, F. J., Stieglitz, T., & Navarro, X. (2005). Long term assessment of axonal regeneration through polyimide regenerative electrodes to interface the peripheral nerve. *Biomaterials, 26*, 2021–2031.

Lago, N., Udina, E., Ramachandran, A., & Navarro, X. (2007). Neurobiological assessment of regenerative electrodes for bidirectional interfacing injured peripheral nerves. *IEEE Transactions on Biomedical Engineering, 54*, 1129–1137.

Lago, N., Yoshida, K., Koch, K. P., & Navarro, X. (2007). Assessment of biocompatibility of chronically implanted polyimide and platinum intrafascicular electrodes. *IEEE Transactions on Biomedical Engineering, 54*, 281–290.

Lawrence, S. M., Dhillon, G. S., Jensen, W., Yoshida, K., & Horch, K. W. (2004). Acute peripheral nerve recording characteristics of polymer-based longitudinal intrafascicular electrodes. *IEEE Transactions on Neural Systems and Rehabilitation Engineering, 12*, 345–348.

Leventhal, D. K., & Durand, D. M. (2003). Subfascicle stimulation selectivity with the flat interface nerve electrode. *Annals of Biomedical Engineering*, *31*, 643–652.

Loeb, G. E., & Peck, R. A. (1996). Cuff electrodes for chronic stimulation and recording of peripheral nerve activity. *Journal of Neuroscience Methods*, *64*, 95–103.

Lotfi, P., Garde, K., Chouhan, A. K., Bengali, E., & Romero-Ortega, M. I. (2011). Modality-specific axonal regeneration: Toward selective regenerative neural interfaces. *Frontiers in Neuroengineering*, *4*, 11.

Lundborg, G., & Kanje, M. (1996). Bioartificial nerve grafts: A prototype. *Scandinavian Journal of Plastic and Reconstructive Surgery and Hand Surgery*, *30*, 105–110.

Lyons, G. M., Sinkjær, T., Burridge, J. H., & Wilcox, D. J. (2002). A review of portable FES-based neural orthoses for the correction of drop foot. *IEEE Transactions on Neural Systems and Rehabilitation Engineering*, *10*, 260–279.

Mayer, R. D., & Howard, F. M. (2008). Sacral nerve stimulation: Neuromodulation for voiding dysfunction and pain. *Neurotherapeutics*, *5*, 107–113.

Mensinger, A. F., Anderson, D. J., Buchko, C. J., Johnson, M. A., Martin, D. C., Tresco, P. A., et al. (2000). Chronic recording of regenerating VIIIth nerve axons with a sieve electrode. *Journal of Neurophysiology*, *83*, 611–615.

Micera, S., & Navarro, X. (2009). Bidirectional interfaces with the peripheral nervous system. *International Review of Neurobiology*, *86*, 23–38.

Micera, S., Navarro, X., Carpaneto, J., Citi, L., Tonet, O., Rossini, P. M., et al. (2008). On the use of longitudinal intrafascicular peripheral interfaces for the control of cybernetic hand prostheses in amputees. *IEEE Transactions on Neural Systems and Rehabilitation Engineering*, *16*, 453–472.

Micera, S., Rossini, P. M., Rigosa, J., Citi, L., Carpaneto, J., Raspopovic, S., et al. (2011). Decoding of grasping information from neural signals recorded using peripheral intrafascicular interfaces. *Journal of Neuroengineering Rehabilitation*, *8*, 53.

Millán, J. D. R., Rupp, R., Müller-Putz, G. R., Murray-Smith, R., Giugliemma, C., Tangermann, M., et al. (2010). Combining brain–computer interfaces and assistive technologies: State-of-the-art and challenges. *Frontiers in Neuroscience*, *4*, 161.

Müller-Putz, G. R., Scherer, R., Pfurtscheller, G., & Rupp, R. (2005). EEG-based neuroprosthesis control: A step towards clinical practice. *Neuroscience Letters*, *382*, 169–174.

Naples, G. G., Mortimer, J. T., & Yuen, T. G. H. (1990). *Overview of peripheral nerve electrode design and implantation*. Englewood Cliffs, NJ: Prentice Hall (pp. 107–145).

Navarro, X., Calvet, S., Butí, M., Gómez, N., Cabruja, E., Garrido, P., et al. (1996). Peripheral nerve regeneration through microelectrode arrays based on silicon technology. *Restorative Neurology and Neuroscience*, *9*, 151–160.

Navarro, X., Calvet, S., Rodriguez, F. J., Stieglitz, T., Blau, C., Buti, M., et al. (1998). Stimulation and recording from regenerated peripheral nerves through polyimide sieve electrodes. *Journal of the Peripheral Nervous System*, *3*, 91–101.

Navarro, X., Krueger, T. B., Lago, N., Micera, S., Stieglitz, T., & Dario, P. (2005). A critical review of interfaces with the peripheral nervous system for the control of neuroprostheses and hybrid bionic systems. *Journal of the Peripheral Nervous System*, *10*, 229–258.

Navarro, X., Lago, N., Vivó, M., Yoshida, K., Koch, K. P., Poppendieck, W., et al. (2007). Neurobiological evaluation of thin-film longitudinal intrafascicular electrodes as a peripheral nerve interface. In *IEEE 10th International Conference on Rehabilitation Robotics* (pp. 643–649).

Navarro, X., Valderrama, E., Stieglitz, T., & Schüttler, M. (2001). Selective fascicular stimulation of the rat sciatic nerve with multipolar polyimide cuff electrodes. *Restorative Neurology and Neuroscience*, *18*, 9–21.

Nordhausen, C. T., Maynard, E. M., & Normann, R. A. (1996). Single unit recording capabilities of a 100 microelectrode array. *Brain Research*, *726*, 129–140.

Panetsos, F., Avendaño, C., Negredo, P., Castro, J., & Bonacasa, V. (2008). Neural prostheses: Electrophysiological and histological evaluation of central nervous system alterations
due to long-term implants of sieve electrodes to peripheral nerves in cats. *IEEE Transactions on Neural Systems and Rehabilitation Engineering, 16*, 223–232.

Park, S. H., & Lee, S. P. (1998). EMG pattern recognition based on artificial intelligence
techniques. *IEEE Transactions on Rehabilitation Engineering, 6*, 400–405.

Peckham, P. H., Keith, M. W., Kilgore, K. L., Grill, J. H., Wuolle, K. S., Thrope, G. B.,
et al. (2001). Efficacy of an implanted neuroprosthesis for restoring hand grasp in
tetraplegia: A multicenter study. *Archives of Physical Medicine and Rehabilitation, 82*,
1380–1388.

Popovic, M. R., Curt, A., Keller, T., & Dietz, V. (2001). Functional electrical stimulation for
grasping and walking: Indications and limitations. *Spinal Cord, 39*, 403–412.

Popovic, M. R., & Keller, T. (2005). Modular transcutaneous functional electrical stimulation system. *Medical Engineering & Physics, 27*, 81–92.

Ragnarsson, K. T. (2007). Functional electrical stimulation after spinal cord injury: Current
use, therapeutic effects and future directions. *Spinal Cord, 46*, 255–274.

Raspopovic, S., Carpaneto, J., Udina, E., Navarro, X., & Micera, S. (2010). On the identification of sensory information from mixed nerves by using single-channel cuff electrodes. *Journal of Neuroengineering Rehabilitation, 7*, 17.

Rosen, J. M., Grosser, M., & Hentz, V. R. (1990). Preliminary experiments in nerve regeneration through laser-drilled holes in silicon chips. *Restorative Neurology and Neuroscience,
2*(2), 89–102.

Rossini, P. M., Micera, S., Benvenuto, A., Carpaneto, J., Cavallo, G., Citi, L., et al. (2010).
Double nerve intraneural interface implant on a human amputee for robotic hand control. *Clinical Neurophysiology, 121*, 777–783.

Rupp, R., & Gerner, H. J. (2004). Neuroprosthetics of the upper extremity—Clinical application in spinal cord injury and future perspectives. *Biomedizinische Technik/Biomedical
Engineering, 49*, 93–98.

Schiefer, M. A., Polasek, K. H., Triolo, R. J., Pinault, G. C. J., & Tyler, D. J. (2010). Selective stimulation of the human femoral nerve with a flat interface nerve electrode. *Journal
of Neural Engineering, 7*, 026006.

Schuind, F., Abramowicz, D., & Schneeberger, S. (2007). Hand transplantation: The state-
of-the-art. *Journal of Hand Surgery, British & European, 32*, 2–17.

Schultz, A. E., & Kuiken, T. A. (2011). Neural interfaces for control of upper limb prostheses: The state of the art and future possibilities. *PM&R, 3*, 55–67.

Scott, R. N., & Parker, P. A. (1988). Myoelectric prostheses: State of the art. *Journal of Medical
Engineering & Technology, 12*, 143–151.

Sensinger, J. W., Schultz, A. E., & Kuiken, T. A. (2009). Examination of force discrimination
in human upper limb amputees with reinnervated limb sensation following peripheral
nerve transfer. *IEEE Transactions on Neural Systems and Rehabilitation Engineering, 17*,
438–444.

Stein, R. B., & Mushahwar, V. (2005). Reanimating limbs after injury or disease. *Trends in
Neuroscience, 28*, 518–524.

Stieglitz, T. (2007). Neural prostheses in clinical practice: Biomedical microsystems in neurological rehabilitation. *Acta Neurochirurgica Supplement, 97*, 411–418.

Tarler, M. D., & Mortimer, J. T. (2004). Selective and independent activation of four motor
fascicles using a four contact nerve-cuff electrode. *IEEE Transactions on Neural Systems
and Rehabilitation Engineering, 12*, 251–257.

Tesfayesus, W., & Durand, D. M. (2006). Blind source separation of neural recordings and
control signals. In *28th Annual International Conference of the IEEE Engineering in Medicine
and Biology Society* (pp. 731–734).

Tyler, D. J., & Durand, D. M. (1997). A slowly penetrating interfascicular nerve electrode for selective activation of peripheral nerves. *IEEE Transactions on Rehabilitation Engineering, 5,* 51–61.

Tyler, D. J., & Durand, D. M. (2002). Functionally selective peripheral nerve stimulation with a flat interface nerve electrode. *IEEE Transactions on Neural Systems and Rehabilitation Engineering, 10,* 294–303.

Tyler, D. J., & Durand, D. M. (2003). Chronic response of the rat sciatic nerve to the flat interface nerve electrode. *Annals of Biomedical Engineering, 31,* 633–642.

Vallès, M., Rodríguez, A., Borau, A., & Mearin, F. (2009). Effect of sacral anterior root stimulator on bowel dysfunction in patients with spinal cord injury. *Diseases of the Colon and Rectum, 52,* 986–992.

Velliste, M., Perel, S., Spalding, M. C., Whitford, A. S., & Schwartz, A. B. (2008). Cortical control of a prosthetic arm for self-feeding. *Nature, 453,* 1098–1101.

Veraart, C., Grill, W. M., & Mortimer, J. T. (1993). Selective control of muscle activation with a multipolar nerve cuff electrode. *IEEE Transactions on Biomedical Engineering, 40,* 640–653.

Walter, J. S., Griffith, P., Sweeney, J., Scarpine, V., Bidnar, M., McLane, J., et al. (1997). Multielectrode nerve cuff stimulation of the median nerve produces selective movements in a raccoon animal model. *Journal of Spinal Cord Medicine, 20,* 233–243.

Warwick, K., Gasson, M., Hutt, B., Goodhew, I., Kyberd, P., Andrews, B., et al. (2003). The application of implant technology for cybernetic systems. *Archives of Neurology, 60,* 1369–1373.

Waters, R. L., McNeal, D. R., Faloon, W., & Clifford, B. (1985). Functional electrical stimulation of the peroneal nerve for hemiplegia: Long-term clinical follow-up. *Journal of Bone and Joint Surgery, 67,* 792–793.

Weir, R., Mitchell, M., Clark, S., Puchhammer, G., Haslinger, M., Grausenburger, R., et al. (2008). The intrinsic hand—a 22 degree-of-freedom artificial hand-wrist replacement. In *Myoelectric Controls/Powered Prosthetics Symposium* (pp. 233–237).

Wieringa, P. A., Wiertz, R. W. F., De Weerd, E., & Rutten, W. L. C. (2010). Bifurcating microchannels as a scaffold to induce separation of regenerating neurites. *Journal of Neural Engineering, 7,* 016001.

Yoshida, K., & Horch, K. (1993). Selective stimulation of peripheral nerve fibers using dual intrafascicular electrodes. *IEEE Transactions on Biomedical Engineering, 40,* 492–494.

Yoshida, K., & Stein, R. B. (1999). Characterization of signals and noise rejection with bipolar longitudinal intrafascicular electrodes. *IEEE Transactions on Biomedical Engineering, 46,* 226–234.

Zhang, D., Peng, Y., Qi, H., Gao, Q., & Zhang, C. (2009). Application of multielectrode array modified with carbon nanotubes to simultaneous amperometric determination of dihydroxybenzene isomers. *Sensors and Actuators B: Chemical, 136,* 113–121.

CHAPTER THREE

The Use of Shock Waves in Peripheral Nerve Regeneration: New Perspectives?

Thomas Hausner[*,†,‡], **Antal Nógrádi**[*,§,1]

[*]Austrian Cluster for Tissue Regeneration and Ludwig Boltzmann Institute for Experimental and Clinical Traumatology at the Research Centre for Trauma of the Austrian Workers' Compensation Board (AUVA), Vienna, Austria
[†]Department for Trauma Surgery and Sports Traumatology, Paracelsus Medical University, Salzburg, Austria
[‡]Department for Surgery, State Hospital Hainburg, Hainburg, Austria
[§]Department of Anatomy, Histology and Embryology, Faculty of Medicine, University of Szeged, Szeged, Hungary
[1]Corresponding author: e-mail address: nogradi.antal@med.u-szeged.hu

Contents

Abstract

Low-energy extracorporeal shock wave treatment (ESWT) is a relatively new therapeutic tool that is widely used for the treatment of epicondylitis and plantar fasciitis and to foster bone and wound healing. Shock waves, sonic pulses with high energy impact, are thought to induce biochemical changes within the targeted tissues through mechanotransduction. The biological effects of ESWT are manifested in improved vascularization, the local release of growth factors, and local anti-inflammatory effects, but the target cells too are influenced.

ESWT appears to have differential effects on peripheral nerves and has been proved to promote axonal regeneration after axotomy. This review discusses the effects of ESWT on intact and injured peripheral nerves and suggests a multiple mechanism of action.

International Review of Neurobiology, Volume 109
ISSN 0074-7742
http://dx.doi.org/10.1016/B978-0-12-420045-6.00003-1

85

1. INTRODUCTION

Shock waves are transient short-term sonic pulses with a high-peak pressure up to 100 Mpa, followed by a negative pressure of about 5–10 MPa. They have rapid rise times of the order of nanoseconds and short pulse durations ranging up to 5 μs. Shock waves are induced electro-hydraulically and then reflected by a focusing device with either parabolic or ellipsoid geometry. The spatial shape of the pressure field depends on the form of this reflector, and shock waves may therefore be applied in a focused or a defocused manner. Moreover, shock waves can be applied extra- or intracorporeally and either low- or high energy levels may be used.

Focused shock waves are used to disintegrate solid aggregations such as kidney stones or solid deposits in tissues (calcified tendons) that usually contain minerals. For these applications, a high energy level is necessary in order to destroy the kidney stones or calcifications. The high energy transmission in cases of focused shock wave treatment necessitates intravenous sedation or even general anesthesia as this procedure is often very painful. Defocused shock waves are administered in soft tissue diseases such as chronic wounds or ulcerations, and recent applications include ischemic heart disease too (Zimpfer et al., 2009). Defocused shock waves display a different shape of acoustic pressure distribution and hence a larger tissue area is affected. Accordingly, defocused low-energy shock wave treatment does not induce pain in most cases.

Although most shock wave treatments are applied extracorporeally (extracorporeal shock wave treatment, ESWT), this treatment does not produce satisfactory results in all cases. In this situation, the use of intracorporeal shock waves may be suggested, for example, endoscopic intracorporeal shock wave lithotripsy for the treatment of bile stones refractive to traditional endoscopic methods (Attila, May, & Kortan, 2008).

Shock wave treatment may also be divided into high- and low-transfer-energy categories. While both treatment modalities are of therapeutic value, high-energy shock wave treatments are typically used for the destruction of solid aggregations inside or outside tissues, whereas low-energy treatment is administered for tissue repair and regeneration (Mittermayer et al., 2012).

The use of shock waves as a therapeutic approach has a relatively short history. Shock wave treatment was first used for the destruction of urinary stones, including those in the kidney, in the 1980s (Chaussy et al., 1982). A decade later, two groups reported successful treatment of calcifying tendinopathies of the shoulder by the disintegration of calcified deposits, and shortly afterward shock wave treatment was introduced into other fields of medicine. ESWT has become a widely utilized therapeutic tool in regenerative medicine in recent years. It is frequently and successfully administered in painful conditions such as humero-radial epicondylitis (tennis elbow), plantar fasciitis, and other pathological conditions affecting bone-related structures. Chronic wounds, ulcerations, and ischemic heart failure have also been successfully targeted by ESWT (Mittermayer et al., 2012; Nishida et al., 2004; Zimpfer et al., 2009).

In this review, we focus on the use and effects of defocused low-energy ESWT in the peripheral nervous system.

 ## 2. FEATURES OF PERIPHERAL NERVE REGENERATION IN RODENTS AND HUMANS: HOW TO SPEED UP SLOW REGENERATION?

Injuries to peripheral nerves are followed by a rapid process of degenerative events called Wallerian degeneration. These events include changes that are effective in the anterograde direction from the injury site, that is, the disconnection of axons from the target organ, for example, the motor endplate, the breakdown of axon and myelin in the distal stump of the injured nerve, and changes that affect the proximal nerve stump (degeneration up to the first Ranvier node) and mainly the cell body retrograde from the injury: chromatolysis (the classical term for the disintegration of the rough endoplasmatic reticulum), dislocation of the nucleus, and shrinkage of the dendritic tree. The degenerative processes are followed by regenerative events, provided that the injured nerve stumps are in close vicinity and the regenerating axons from the proximal stump are able to enter the vacated endoneural sheaths in the distal stump. The regrowth of axons is supported by the rapid proliferation of Schwann cells in the distal stump providing a contact guide for them. The proliferating Schwann cells align to form the bands of Büngner and restructure the extracellular matrix that contains growth-promoting molecules such as laminin and fibronectin. The growth cone of the regenerating axons actively synthesizes transmembrane integrin molecules, for example, integrin-type alpha5-beta1, which interacts with fibronectin, thereby ensuring the axonal growth. In this way, all the conditions required for the successful regeneration and reinnervation of the targets are provided.

While axons in the rodent peripheral nervous system regenerate at a speed of 2–3 mm/day, so that relatively short distances are rapidly bridged by growing axons, human peripheral nerve injuries are followed by a slower rate of regeneration (1 mm/day). Given the fact that some motor and sensory axons projecting into the lower limb may reach or exceed a length of 1 m, the regeneration and reinnervation of peripheral targets may clearly be an extremely slow process. The slow regeneration in human peripheral nerves is further hampered by the predegenerative process occurring in the distal parts of the nerve to be occupied by regenerating fibers (Gordon, 2010; Gordon et al., 2009).

In view of the long recovery and rehabilitation process after injuries to peripheral nerves, there is a great need for the development of procedures that promote peripheral nerve regeneration in humans and thereby decrease the related social and health-care costs. While the effects of shock waves on wound healing, (Schaden et al., 2007), bone regeneration (Ogden, Alvarez, Levitt, Cross, & Marlow, 2001), and the integration of skin grafts (Kuo et al., 2009; Stojadinovic et al., 2008) have been extensively studied, very little is known as concerns its effects on peripheral nerve regeneration (Hausner et al., 2012; Wu, Lun, Chen, & Chong, 2007).

3. PRESUMED BIOLOGICAL EFFECTS OF ESWT

Shock waves are mechanical events that can stimulate tissues and especially cells. The conversion of physical forces into biochemical signals is a fundamental process required for the development and the physiology of organisms. This process is referred to as mechanotransduction. Physical forces exert a direct influence on protein folding, and force-induced effects on the three-dimensional structures of proteins are therefore involved in a general mechanism through which the activities of enzymes or the interactions between proteins may lead to signal modification (Orr, Helmke, Blackmann, & Schwartz, 2006). The manner in which ESWT-induced mechanotransduction is manifested in target cells and tissues is still not clear. There are a number of proved facts or theories concerning the cascade of actions stemming from shock wave treatment and resulting in angiogenesis or neovascularization (Sadoun and Reed, 2003; Stojadinovic et al., 2008; Wang et al., 2004), anti-inflammatory effects (Davis et al., 2009), the release of growth factors (Hausdorf et al., 2011), and the activation of progenitor cells and stem cells (Mittermayer et al., 2012; Sadoun and Reed, 2003).

There are various mechanisms behind the effects in tissues treated with shock waves. It has been reported that angiogenesis is induced by increased levels of vascular endothelial growth factor-A, which in turn is triggered by upregulated activities of nitric oxide synthase (NOS), and extracellular signal-regulated kinase. On the other hand, enhanced NOS activity also appears to be responsible for the activation of hypoxia-inducible factor-1 in a variety of cells, depending on the target of ESWT. Low-energy shock wave treatment has likewise proved to be effective in downregulating immune responses in acute wounds. ESWT has been reported to reduce the invasion of macrophages and polymorphonuclear leucocytes into the wound area, together with the suppressed production of proinflammatory cytokines and chemokines at the wound matrix (Davis et al., 2009; Kuo et al., 2009). Similar to its role in inducing angiogenesis in shock wave-treated tissues, the regulatory function of NOS has been suggested in the downregulation of inflammatory events in these conditions (Fig. 3.1; Mariotto et al., 2009). Others have described the increased release of fibro-blast growth factor-2, acting on osteoblasts (Hausdorf et al., 2011), while osteocalcin, a major bone protein playing an important role in bone miner-alization, is reportedly upregulated in regenerating the bone after ESWT (Martini et al., 2003). In contrast with these molecules, the role of trans-forming growth factor-beta remains controversial (Hausdorf et al., 2011; Martini et al., 2003).

It has been suggested that mesenchymal stem cells may differentiate toward tissue-specific progenitor cells such as osteoblasts in response to ESWT (Chen et al., 2004), and the moderate recruitment of endothelial progenitor cells has been described (Tinazzi et al., 2011). However, the extent to which these mechanisms are able to contribute to the tissue repair following ESWT is not clear at present.

4. EFFECTS OF ESWT ON PERIPHERAL NERVES

4.1. Effects of ESWT on sensory nerves

Shock waves have been used extensively to study their effects on sensory nerves and nerve endings. Application of 1000 impulses of shock waves (0.08 mJ/mm, 2.4 Hz) resulted in the degeneration of sensory nerve fibers and endings followed by reinnervation of the affected skin areas (Ohtori et al., 2001). These changes were accompanied by the reversible and rapid loss of the immunohistochemical markers protein gene product 9.5 and

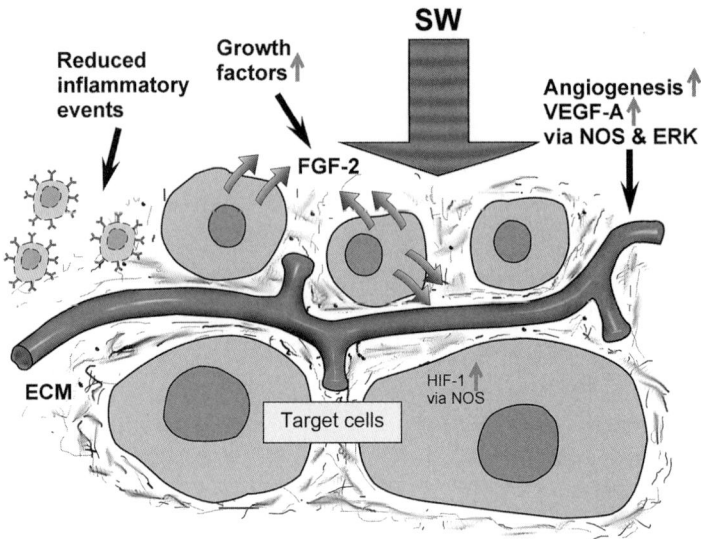

Figure 3.1 Schematic drawing depicting the sites of action by shock waves in various tissues (with the exception of peripheral nerves). The target cells of shock wave treatment are embedded in the extracellular matrix, surrounded by various other cell types, including resident and invading mononuclear and polymorphonuclear immune cells. ESWT has been proved to induce the release of growth factors (e.g., FGF-2) from the cells surrounding the target cells, to improve angiogenesis within the tissues, and to reduce the secretion of inflammatory cytokines and the invasion of immune cells. On the other hand, tissue-specific target cells are known to secrete factors such as hypoxia-inducible factor-1 (HIF-1). Several of these processes are regulated via the activation of nitric oxide synthase (NOS); it should be noted that the extent to which these processes are induced varies with the type of tissue (ECM, extracellular matrix; ERK, extracellular signal-regulated kinase; FGF-2, fibroblast growth factor-2; VEGF-A, vascular endothelial growth factor-A; SW, shock waves). (For color version of this figure, the reader is referred to the online version of this chapter.)

calcitonin gene–related peptide. However, a second application of the same dose of shock waves had a cumulative effect on the treated nerves, leading to delayed reinnervation (Takahashi, Ohtori, Saisu, Moriya, & Wada, 2006). It appears, therefore, that shock wave-treated nerves develop a "memory effect" after the first treatment, and ESWT repeated shortly after the first treatment is not beneficial. It is expected that ESWT induces subtle changes in the affected neurones whose axons have been treated. Murata et al. (2006) detected an increased expression of activating transcription factor 3 (ATF-3) and growth–associated phosphoprotein 43 (GAP-43) in dorsal root ganglion neurones of shock wave-treated rats, indicating that the molecular changes after ESWT are not restricted to the treated axons: their cell bodies are also activated in a retrograde manner. The question remains open as to whether doses of ESWT in the therapeutic range would induce similar changes as the 2000 impulses applied in this study. ATF-3 and GAP-43 are markers thought to be associated with the activation of neurones and glial cells (Schwann cells) after peripheral nerve injuries (Hunt et al., 2004; Saito and Dahlin, 2008).

As regards the dose–effect relationship of ESWT on peripheral nerves, a large body of evidence suggests that shock wave doses greater than 900 impulses combined with a flux density of 0.08 mJ/mm^2 induce damage to the affected nerves, manifested in impaired electrophysiological conduction parameters (Wu et al., 2007), a disrupted neurofilament staining pattern of the treated axons (Hausner et al., 2012), and degeneration of the myelin sheaths at the levels of light and electron microscopy (Bolt et al., 2004). These doses appeared to damage motor and sensory nerves equally (Bolt et al., 2004; Wu et al., 2007). Our experimental and clinical experience indicates that the therapeutically applicable dose for the promotion of nerve regeneration without side effects is likely to be lower than 500 impulses (0.1 mJ/mm^2, 4 Hz) (Hausner et al., 2012). The effect of such doses is highly dependent on the depth of the target tissue and the treated surface area.

4.2. Effects of ESWT on motor nerves

The question arose of whether doses of shock wave treatment that did not cause degenerative events in the affected peripheral nerve segments would foster the regeneration of injured axons in a rodent model. It was clearly demonstrated that ESWT applied at a dose of 300 impulses and 0.1 mJ/mm^2 did not induce the disintegration of neurofilaments within the axons of the sciatic nerve (Hausner et al., 2012). The efficacy of this ESWT scheme was tested in an autologous rat sciatic nerve model, where an 8-mm long autograft was excised and coapted with the proximal and distal stumps. When shock wave treatment was applied immediately after surgical reconstruction, a significantly improved rate of axonal regeneration was observed as early as 3 weeks after the injury. Not only were more regenerating axons found in the reinnervated distal stump of the shock wave-treated nerves, but also this early reinnervation was accompanied by moderate values of axon conduction beyond the distal coaptation site. The morphological and functional reinnervation of the denervated hind limb muscles could be expected only at later time points. Functional tests revealed a clear improvement in the ESWT animals from 4 weeks onward, but this difference in improved locomotor pattern was no longer detectable from week 10 after surgery. Twelve weeks after injury, none of the morphological, functional, or electrophysiological parameters indicated differences between the treated and the untreated animals, with the exception of the conduction velocity, which was still significantly higher in the ESWT group (Fig. 3.2).

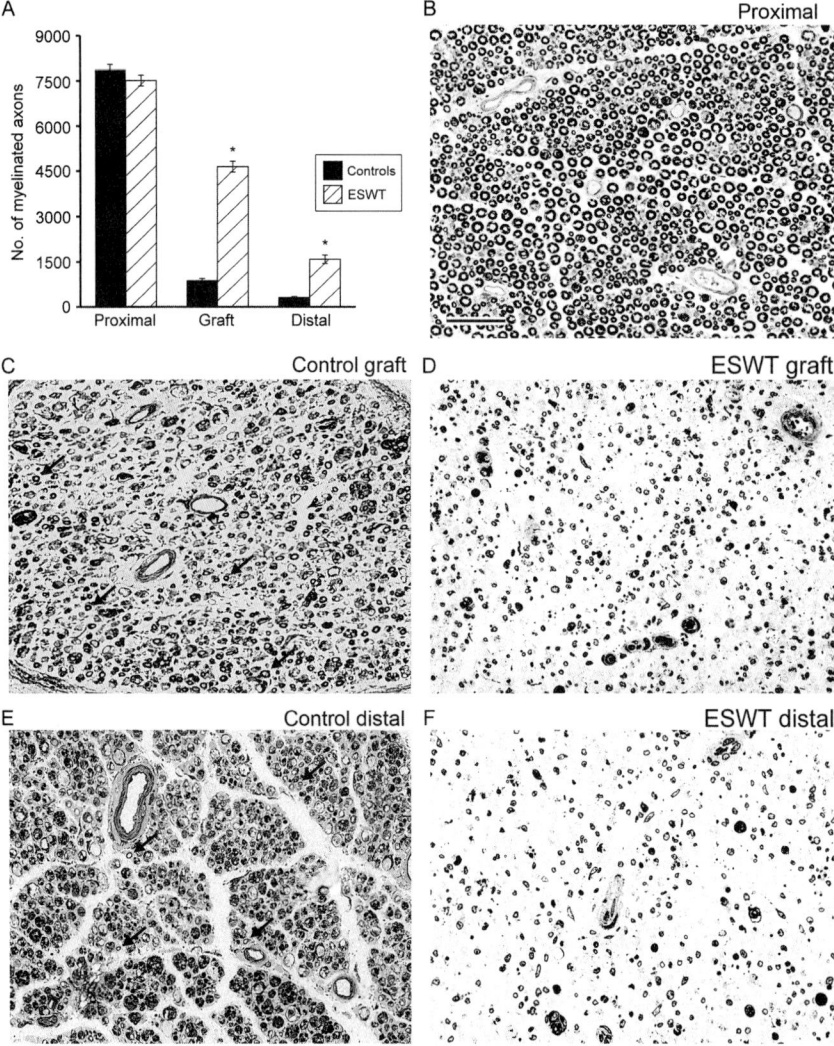

Figure 3.2 Axonal regeneration in control and extracorporeal shock wave-treated (ESWT) peripheral nerves 3 weeks and 3 months after surgery. (A) The columns show the numbers of myelinated fibers found in the middle of the graft and 2 mm proximal and distal to the graft in ESWT and control animals 3 weeks after axotomy (left). The

Ultrastructural analysis of the nerve grafts 3 weeks after the injury revealed that not only were there more regenerated and well-myelinated axons in the ESWT nerves, but also the endoneurium was free from reactive cells and degenerated myelin profiles, which were present in abundance in the untreated nerves (Fig. 3.3; Hausner et al., 2012). These findings indicated that the improved rate of axonal regeneration and the clearing-up of the degenerated structures in the denervated nerves are strongly related. It remains for future studies to establish whether either of these processes enjoys priority over the other in the temporal sequence of events.

5. CONCLUSION

Shock waves were introduced into the arsenal of modern human medical therapy some 30 years ago (Shrivastava and Kailash, 2005; Thiel, 2001). Following the initial treatment trials on urolithiasis, extracorporeal shock waves were introduced both preclinically and clinically for the treatment of acute and chronic soft and hard tissue healing problems (Ogden et al., 2001). In most cases, improvements in the soft and hard tissue healing processes were found to be associated with increased levels of vascularization, and this mechanism of action was therefore considered to be a general, but not overall scenario for shock wave-induced improvement (Wang et al., 2004; Yan, Zeng, Chai, Luo, & Li, 2008; Zimpfer et al., 2009). However, it has subsequently been demonstrated that other nonvascular mechanisms contribute to the tissue repair (for details, see Fig. 3.1).

numbers of myelinated axons in the graft and distal to the grafting site are much higher in the ESWT animals than in the controls. There was no significant difference between the controls and the ESWT animals in the numbers of myelinated axons distal to the graft 3 months after axotomy and grafting (right). *Significant difference ($p < 0.05$) between the control and the ESWT groups by ANOVA, computed by using Tukey's all pairwise multiple comparison procedures. (B–F) Photographs of semithin cross-sections from the proximal stump (B), the middle of the graft (C, D), and the distal stump (E, F) 3 weeks after axotomy. The shock wave-treated peripheral nerves (ESWT) contain more myelinated axons, while the control nerves display far fewer regenerated axons (arrows) and are full of degenerated myelin sheaths and reactive cells. (G, H): Photographs of semithin cross-sections from the distal stump 3 months after axotomy. There is no striking difference between the ESWT and control nerves, although the myelin sheaths of the regenerated axons appear thinner than those seen in the intact proximal stump (B). Methylene blue–thionin staining according to Rüdeberg, scale bar = 25 μm. *This figure is reproduced from the publication by Hausner et al. (2012), with the kind permission of Elsevier/Rightslink.*

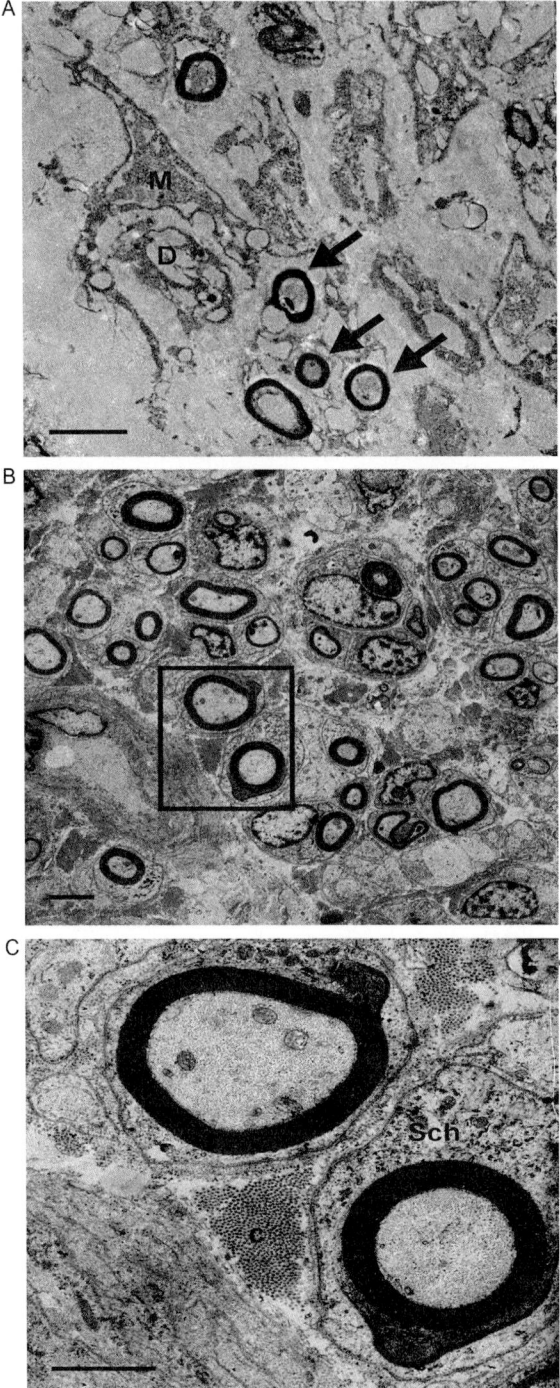

Figure 3.3 Electron microscopic photographs of control (A) and shock wave-treated peripheral nerves 3 weeks after surgery. Panel (A) shows several degenerated myelin sheaths (D) engulfed by macrophages (M). A few myelinated regenerated axons (arrows) too can be seen. In panel (B), a high number of myelinated axons are present

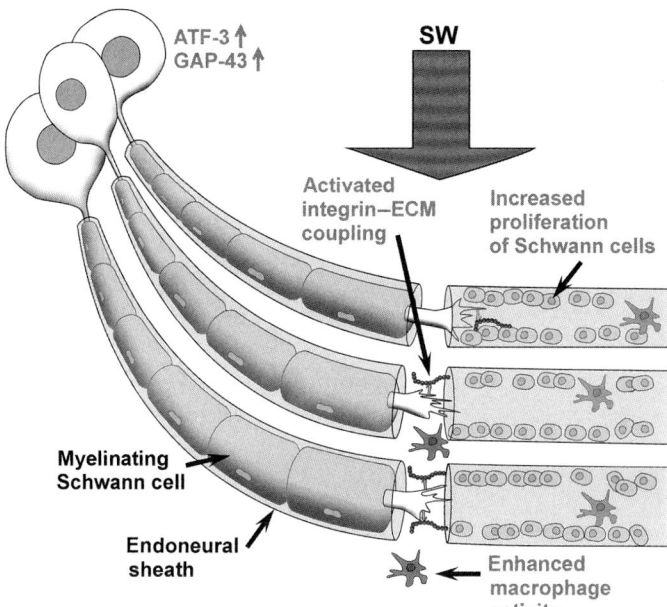

Figure 3.4 Schematic drawing displaying the possible sites of action of ESWT in a regenerating peripheral nerve and related cell bodies. In an untreated peripheral nerve, regenerating neurites enter the vacated endoneural sheaths of the degenerated distal peripheral nerve stump and grow along the aligned proliferating Schwann cells (bands of Büngner), provided that the proximal and distal stumps are sufficiently close to each other. It is suggested here that ESWT may improve the rate of axonal regeneration through the activation of integrin molecules expressed on the axonal growth cones, thereby promoting stronger binding to the various extracellular molecules, such as laminin and fibronectin. The rate of proliferation of Schwann cells in the distal stump may also increase, in conjunction with a more pronounced macrophage activity, which results in the faster and more effective clearance of myelin debris. These actions may have a cumulative effect on nerve regeneration, leading to a faster and more accurate reinnervation process. The changes in the distal stump of the nerve and the more effective axonal regeneration is accompanied by molecular changes in the related cell bodies, that is, upregulation of transcription factor ATF-3 and GAP-43. Apart from these actions marked in red, it appears conceivable that many other molecular mechanisms too are upregulated in order to support extensive axonal growth. The term endoneural sheath refers to the fine network of reticular fibers and extracellular matrix molecules around each myelinated axon as part of the endoneurium (ATF-3, activating transcription factor-3; ECM, extracellular matrix; GAP-43, growth-associated phosphoprotein-43; SW, shock waves). (For interpretation of the references to color in this figure legend, the reader is referred to the online version of this chapter.)

without reactive cells, but surrounded by Schwann cells. Panel (C) presents a higher magnification of the framed area in (B). Note the remyelinating Schwann cells (Sch) and some collagen bundles (C) in the endoneurium. Scale bar in (A) and (B) = 2 µm, in (C) = 1 µm. *This figure is reproduced from the publication by Hausner et al. (2012), with the kind permission of Elsevier/Rightslink.*

The present review has surveyed the findings that describe the effects of shock waves on intact and injured peripheral nerves. Although only limited information is available on the mechanism of action of shock wave treatment on peripheral nerves, improved vascularization does not appear to play a direct role in promoting axon regeneration after axotomy. Axonal regeneration in the peripheral nerves is known to be promoted by several cellular and molecular components of the nerve, including the coupling of integrin molecules situated on the axonal growth cone membrane with the abundant extracellular molecules (Lefcort, Venstrom, McDonald, & Reichardt, 1992; Low, Nógrádi, Vrbová, & Greensmith, 2003; Tomaselli et al., 1993), the proliferation of activated Schwann cells in the degenerated distal stump of the nerve (Stoll and Müller, 1999), and the clear role played by activated macrophages (Dailey, Avellino, Benthem, Silver, & Kliot, 1998; Horie et al., 2004; Hughes and Perry, 2000) in the removal of myelin debris (Fig. 3.4). We therefore suggest that ESWT may augment and potentiate the mechanisms described earlier in a regenerating peripheral nerve segment. It is to be expected that ESWT will become more widely used in the treatment of injuries and pathological conditions affecting peripheral nerves.

ACKNOWLEDGMENTS

The authors are indebted to the Lorenz Böhler Fonds for financial support. The excellent artwork of Mr. Gábor Márton is gratefully acknowledged. We thank Dr. David Durham for a critical reading of the chapter.

REFERENCES

Attila, T., May, G. R., & Kortan, P. (2008). Nonsurgical management of an impacted mechanical lithotriptor with fractured traction wires: Endoscopic intracorporeal electro-hydraulic shock wave lithotripsy followed by extra-endoscopic mechanical lithotripsy. *Canadian Journal of Gastroenterology, 22*(8), 699–702.

Bolt, D. M., Burba, D. J., Hubert, J. D., Strain, G. M., Hosgood, G. L., Henk, W. G., et al. (2004). Determination of functional and morphologic changes in palmar digital nerves after nonfocused extracorporeal shock wave treatment in horses. *American Journal of Veterinary Research, 65*(12), 1714–1718.

Chaussy, C., Schmiedt, E., Jocham, D., Brendel, W., Forssmann, B., & Walter, V. (1982). First clinical experience with extracorporeally induced destruction of kidney stones by shock waves. *The Journal of Urology, 127*, 417.

Chen, Y. J., Wurtz, T., Wang, C. J., Kuo, Y. R., Yang, K. D., Huang, H. C., et al. (2004). Recruitment of mesenchymal stem cells and expression of TGF-beta 1 and VEGF in the early stage of shock wave-promoted bone regeneration of segmental defect in rats. *Journal of Orthopaedic Research, 22*(3), 526–534.

Dailey, A. T., Avellino, A. M., Benthem, L., Silver, J., & Kliot, M. (1998). Complement depletion reduces macrophage infiltration and activation during Wallerian degeneration and axonal regeneration. *The Journal of Neuroscience, 18*(17), 6713–6722.

Davis, T. A., Stojadinovic, A., Anam, K., Amare, M., Naik, S., Peoples, G. E., et al. (2009). Extracorporeal shock wave therapy suppresses the early proinflammatory immune response to a severe cutaneous burn injury. *International Wound Journal*, *6*, 11–21.

Gordon, T. (2010). The physiology of neural injury and regeneration: The role of neurotrophic factors. *Journal of Communication Disorders*, *43*(4), 265–273.

Gordon, T., Chan, K. M., Sulaiman, O. A., Udina, E., Amirjani, N., & Brushart, T. M. (2009). Accelerating axon growth to overcome limitations in functional recovery after peripheral nerve injury. *Neurosurgery*, *65*(4), A132–A144.

Hausdorf, J., Sievers, B., Schmitt-Sody, M., Jansson, V., Maier, M., & Mayer-Wagner, S. (2011). Stimulation of bone growth factor synthesis in human osteoblasts and fibroblasts after extracorporeal shock wave application. *Archives of Orthopaedic and Trauma Surgery*, *131*(3), 303–309.

Hausner, T., Pajer, K., Halat, G., Hopf, R., Schmidhammer, R., Redl, H., et al. (2012). Improved rate of peripheral nerve regeneration induced by extracorporeal shock wave treatment in the rat. *Experimental Neurology*, *236*(2), 363–370. http://dx.doi.org/10.1016/j.expneurol.2012.04.019, Epub May 1, 2012.

Horie, H., Kadoya, T., Hikawa, N., Sango, K., Inoue, H., Takeshita, K., et al. (2004). Oxidized galectin-1 stimulates macrophages to promote axonal regeneration in peripheral nerves after axotomy. *The Journal of Neuroscience*, *24*(8), 1873–1880.

Hughes, P. M., & Perry, V. H. (2000). The role of macrophages in degeneration and regeneration in the peripheral nervous system. In N. Saunders, K. M. Dzieqielewska, & N. R. Saunders (Eds.), *Degeneration and regeneration in the nervous system* (pp. 263–283). The Netherlands: Harwood Academic.

Hunt, D., Hossain-Ibrahim, K., Mason, M. R., Coffin, R. S., Lieberman, A. R., Winterbottom, J., et al. (2004). ATF3 upregulation in glia during Wallerian degeneration: Differential expression in peripheral nerves and CNS white matter. *BMC Neuroscience*, *4*(5), 9.

Kuo, Y. R., Wang, C. T., Wang, F. S., Yang, K. D., Chiang, Y. C., & Wang, C. J. (2009). Extracorporeal shock wave treatment modulates skin fibroblast recruitment and leukocyte infiltration for enhancing extended skin-flap survival. *Wound Repair and Regeneration*, *17*, 80–87.

Lefcort, F., Venstrom, K., McDonald, J. A., & Reichardt, L. F. (1992). Regulation of expression of fibronectin and its receptor, alpha 5 beta 1, during development and regeneration of peripheral nerve. *Development*, *116*(3), 767–782.

Low, H. L., Nógrádi, A., Vrbová, G., & Greensmith, L. (2003). Axotomized motoneurons can be rescued from cell death by peripheral nerve grafts: The effect of donor age. *Journal of Neuropathology and Experimental Neurology*, *62*(1), 75–87.

Mariotto, S., Carcereri de Prati, A., Cavalieri, E., Amelio, E., Marlinghaus, E., & Suzuki, H. (2009). Extracorporeal shock wave therapy in inflammatory diseases: Molecular mechanism that triggers anti-inflammatory action. *Current Medicinal Chemistry*, *16*, 2366–2372.

Martini, L., Giavaresi, G., Fini, M., Torricelli, P., de Pretto, M., Schaden, W., et al. (2003). Effect of extracorporeal shock wave therapy on osteoblast like cells. *Clinical Orthopaedics and Related Research*, *413*, 269–280.

Mittermayer, R., Antonic, V., Hartinger, J., Kaufmann, H., Redl, H., Téot, L., et al. (2012). Extracorporeal shock wave therapy (ESWT) for wound healing: Technology, mechanisms, and clinical efficacy. *Wound Repair and Regeneration*, *20*, 456–465.

Murata, R., Ohtori, S., Ochiai, N., Takahashi, N., Saisu, T., Moriya, H., et al. (2006). Extracorporeal shockwaves induce the expression of ATF3 and GAP-43 in rat dorsal root ganglion neurons. *Autonomic Neuroscience*, *128*(1–2), 96–100.

Nishida, T., Shimokawa, H., Oi, K., Tatewaki, H., Uwatoku, T., Abe, K., et al. (2004). Extracoporeal cardiac shock wave therapy ameliorates markedly ameliorates ischemia-induced myocardial dysfunction pigs in vivo. *Circulation*, *110*, 3055.

Ogden, J. A., Alvarez, R., Levitt, R., Cross, G. L., & Marlow, M. (2001). Shockwave therapy for chronic proximal plantar fasciitis. *Clinical Orthopaedics and Related Research, 387,*
47–59.
Ohtori, S., Inoue, G., Mannoji, C., Saisu, T., Takahashi, K., Mitsuhashi, S., et al. (2001).
Shock wave application to rat skin induces degeneration and reinnervation of sensory
nerve fibres. *Neuroscience Letters, 315*(1–2), 57–60.
Orr, A. W., Helmke, B. P., Blackmann, B. R., & Schwartz, M. A. (2006). Mechanisms of
mechanotransduction. *Developmental Cell, 10,* 11–20.
Sadoun, E., & Reed, M. J. (2003). Impaired angiogenesis in aging is associated with alter-
ations in vessel density, matrix composition, inflammatory response, and growth factor
expression. *The Journal of Histochemistry and Cytochemistry, 51,* 1119–1130.
Saito, H., & Dahlin, L. B. (2008). Expression of ATF3 and axonal outgrowth are impaired
after delayed nerve repair. *BMC Neuroscience, 9,* 88.
Schaden, W., Thiele, R., Kolpl, C., Pusch, M., Nissan, A., Attinger, C. E., et al. (2007).
Shock wave therapy for acute and chronic soft tissue wounds: A feasibility study. *The
Journal of Surgical Research, 143,* 1–12.
Shrivastava, S. K., & Kailash, S. K. (2005). Shock wave treatment in medicine. *Journal of Bio-
sciences, 30*(2), 269–275.
Stojadinovic, A., Elster, E. A., Anam, K., Tadaki, D., Amare, M., Zins, S., et al. (2008).
Angiogenic response to extracorporeal shock wave treatment in murine skin isografts.
Angiogenesis, 11, 369–380.
Stoll, G., & Müller, H. W. (1999). Nerve injury, axonal degeneration and neural regener-
ation: Basic insights. *Brain Pathology, 9*(2), 313–325.
Takahashi, N., Ohtori, S., Saisu, T., Moriya, H., & Wada, Y. (2006). Second application of
low-energy shock waves has a cumulative effect on free nerve endings. *Clinical Orthopae-
dics and Related Research, 443,* 315–319.
Thiel, M. (2001). Application of shock waves in medicine. *Clinical Orthopaedics and Related
Research, 387,* 18–21.
Tinazzi, E., Amelio, E., Marangoni, E., Guerra, C., Puccetti, A., Codella, O. M., et al.
(2011). Effects of shock wave therapy in the skin of patients with progressive systemic
sclerosis: A pilot study. *Rheumatology International, 31*(5), 651–656.
Tomaselli, K. J., Doherty, P., Emmett, C. J., Damsky, C. H., Walsh, F. S., & Reichardt, L. F.
(1993). Expression of beta 1 integrins in sensory neurons of the dorsal root ganglion and
their functions in neurite outgrowth on two laminin isoforms. *The Journal of Neuroscience,
13*(11), 4880–4888.
Wang, F. S., Wang, C. J., Chen, Y. J., Chang, P. R., Huang, Y. T., Sun, Y. C., et al. (2004).
Ras induction of superoxide activates ERK-dependent angiogenic transcription factor
HIF-1alpha and VEGF-A expression in shock wave-stimulated osteoblasts. *The Journal
of Biological Chemistry, 279,* 10331–10337.
Wu, Y. H., Lun, J. J., Chen, W. S., & Chong, F. C. (2007). The electrophysiological and
functional effect of shock wave on peripheral nerves. *Conference Proceedings—IEEE
Engineering in Medicine and Biology Society,* 2369–2372.
Yan, X., Zeng, B., Chai, Y., Luo, C., & Li, X. (2008). Improvement of blood flow, expression of nitric oxide, and vascular endothelial growth factor by low-energy shockwave
therapy in random-pattern skin flap model. *Annals of Plastic Surgery, 61,* 646–653.
Zimpfer, D., Aharinejad, S., Holfeld, J., Thomas, A., Dumfarth, J., Rosenhek, R., et al.
(2009). Direct epicardial shock wave therapy improves ventricular function and induces
angiogenesis in ischemic heart failure. *The Journal of Thoracic and Cardiovascular Surgery,
137,* 963–970.

CHAPTER FOUR

Phototherapy and Nerve Injury: Focus on Muscle Response ☆

Shimon Rochkind*,†,1, Stefano Geuna‡, Asher Shainberg*

*Faculty of Life Science, Bar-Ilan University, Ramat-Gan, Israel
†Division of Peripheral Nerve Reconstruction, Department of Neurosurgery, Tel Aviv Sourasky Medical Center, Tel Aviv University, Tel Aviv, Israel
‡Department of Clinical and Biological Sciences, University of Turin, Turin, Italy
1Corresponding author: e-mail address: rochkind@zahav.net.il

Contents

Abstract

Preservation of biochemical processes in muscles is a major challenge in patients with severe peripheral nerve injury. In this chapter, we address the effects of laser irradiation and biochemical transformation in muscle, using *in vitro* and *in vivo* experimental models. The authors attempt to explain the possible mechanism of laser phototherapy applied on skeletal muscle on the basis of literature review and new results. A detailed knowledge of the evolution of endplates acetylcholine receptors and creatine kinase activity following laser irradiation can help to understand the therapeutic effect of laser phototherapy on muscle. This study showed that the laser phototherapy increases biochemical activity in intact muscle and thus could have direct therapeutic applications on muscle, especially during progressive atrophy resulting from peripheral nerve injury.

1. INTRODUCTION

Posttraumatic muscle denervation is a common and disabling outcome of various types of accidents which often involve young people (Campbell, 2008). Time to reinnervation is one of the most important

---

☆Part of this work is taken from Dr. Shimon Rochkind's Ph.D. dissertation

International Review of Neurobiology, Volume 109
ISSN 0074-7742
http://dx.doi.org/10.1016/B978-0-12-420045-6.00004-3

DNA synthesis was determined in the cells by measuring the incorporation of ^{3}H-thymidine for 2–3 h in the growth medium (0.5 µCi/ml).

Laser irradiation was applied on the cells directly using a 632.8-nm, 5-mW
He–Ne laser, and diameter of laser beam 2.7 mm × 2.7 mm. Level measurements of DNA and CK in young and mature skeletal muscle cells were performed at seven time points: control (for young and mature cells,
respectively) and 1, 3, 7, 10, 14, and 21 min of laser irradiation. Total
CK activity was measured in the cell homogenate.

Statistical analysis and calculations were done using MatLab software
(Ver. 2008b, The MathWorks, Inc.). We used nonparametric statistics since
the number of rats in each sample time point was too small to evaluate normal deviation. The figure presentation is aligned with our statistics, thus
all figures are presented with Median ± Mad. All significance levels were
calculated using a Mann–Whitney U test when samples are independent
observations (e.g., when comparing the results between rats that were
treated with laser and ones that had no treatment) and Wilcoxon signed-rank
test when comparing two related samples (e.g., when comparing the results
between laser-irradiated and nonirradiated leg of the same rat). Correlations
between muscles in the same rat (laser-irradiated and nonirradiated) were
calculated to evaluate the dynamic change of AChR and CK in time.

In vivo muscle response. Figures 4.1 and 4.2 present qualitative changes
in amount of AChR and CK activity in irradiated (first 14 days) and nonirradiated intact gastrocnemius muscles. Laser irradiation significantly
increased CK activity ($p < 0.05$) and AChR level ($p < 0.01$) in time
periods of 30–60 days in comparison with the nonirradiated gastrocnemius
muscles.

In vitro muscle cells response. Figure 4.3 shows the effect of 632-nm laser
irradiation on the synthesis of DNA in young and mature skeletal muscle
cells at seven time points (control, 1, 3, 7, 10, 14, 21 min of irradiation).
Similar increase in DNA synthesis in both groups was found with tendency
to decrease during the radiated time. Figure 4.4 shows the effect of 632-nm
laser irradiation on the CK activity in young and mature skeletal muscle cells
at seven time points (control [4;4 for young and mature cells, respectively];
min 1 [$n = 0;4$]; min 3 [$n = 4;4$]; min 7 [$n = 3;4$]; min 10 [$n = 0;4$]; min
14 [$n = 3;2$]; min 21 [$n = 3;3$]). In the laser-irradiated mature muscle fibers,
the activity of CK was more expressed than in young cells and control level.
Figure 4.5 shows in comparison the effect of 632-nm laser irradiation on
CK activity and DNA synthesis in young and mature skeletal muscle cells.
To evaluate the laser treatment effect, we calculated all cell results between 1

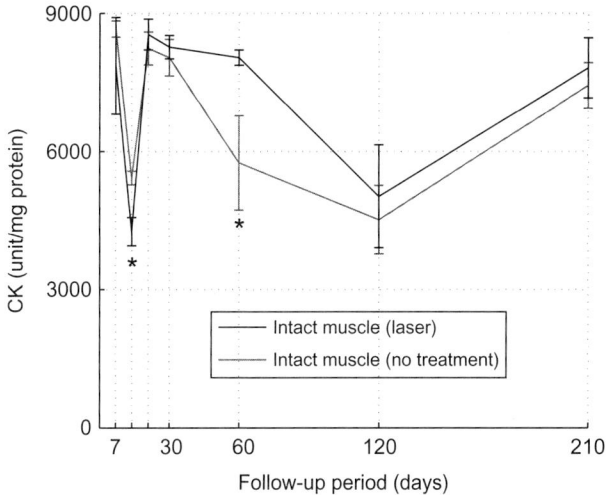

Figure 4.1 Longitudinal level measurements of CK levels in intact gastrocnemius muscle in rats that underwent laser irradiation (black lines) and rats that had no additional treatment (gray lines). Statistically significant change in CK levels between laser-irradiated intact muscles and nonirradiated intact muscles is presented with black asterisks. Asterisks denote significant p-values ($^*p < 0.05$; $^{**}p < 0.01$).

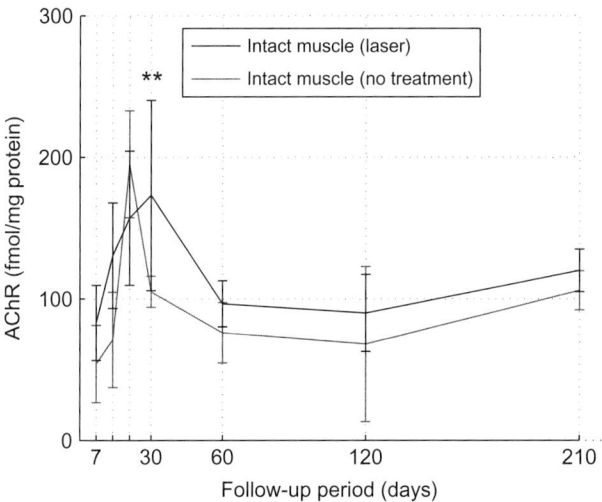

Figure 4.2 Longitudinal level measurements of AChR levels in intact gastrocnemius muscle in rats that underwent laser irradiation (black lines) and rats that had no additional treatment (gray lines). Statistically significant change in AChR levels between laser-irradiated intact muscles and nonirradiated intact muscles is presented with black asterisks. Asterisks denote significant p-values ($^*p < 0.05$; $^{**}p < 0.01$).

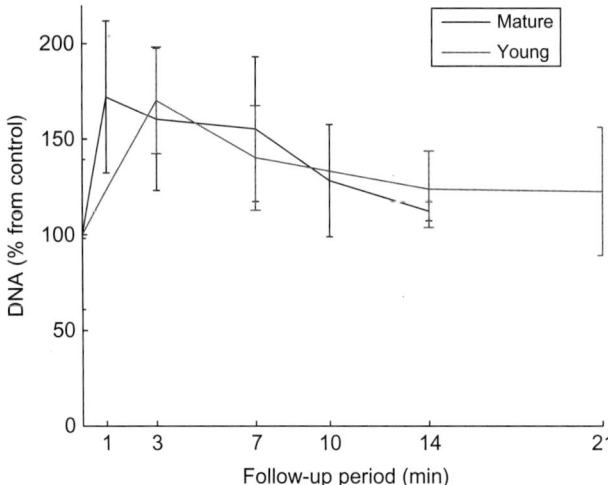

Figure 4.3 Effect of 632-nm laser irradiation on the synthesis of DNA in young and mature skeletal muscle cells. Longitudinal level measurements of DNA synthesis that underwent laser treatment at seven time points (control for young and mature cells, respectively), 1, 3, 7, 10, 14, and 21 min.

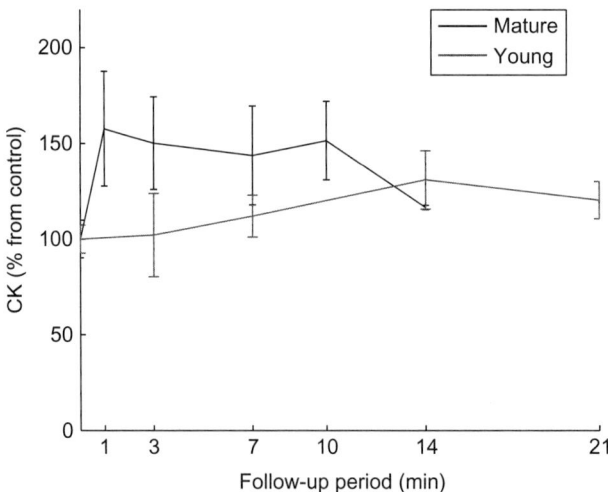

Figure 4.4 Effect of 632-nm laser irradiation on CK activity in young and mature skeletal muscle fibers. Longitudinal level measurements of CK level that underwent laser irradiation at seven time points: control (for young and mature cells, respectively), 1, 3, 7, 10, 14, and 21 min.

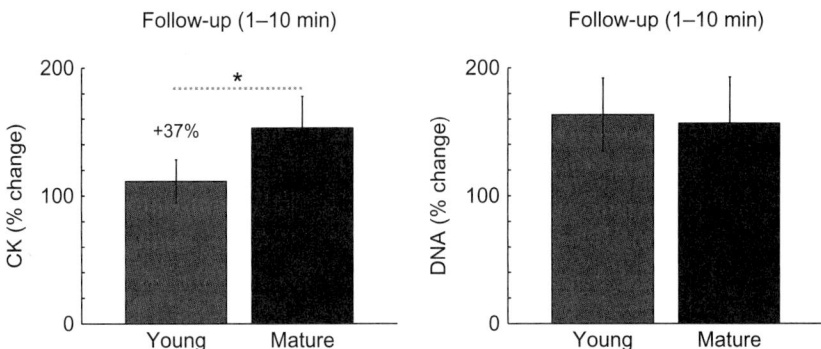

Figure 4.5 Comparative effect of 632-nm laser irradiation on CK activity and DNA synthesis in young and mature skeletal muscle cells. To evaluate the laser treatment effect, we calculated all cells results between 1 and 10 min as there is no trend of change in these time points. Asterisks denote significant p-values ($^{*}p < 0.05$).

and 10 min as there were no trend change in these time points. The laser effect on CK activity increased in the mature cells in comparison with young cells ($p < 0.05$).

3. DISCUSSION AND FUTURE PERSPECTIVES

In this study, we have investigated the influence of low-power laser irradiation on CK activity and the amount AChRs in intact gastrocnemius muscle (*in vivo*), and on the synthesis of DNA and CK in muscle cells (*in vitro*) in order to estimate biochemical transformation on cellular and biochemical levels.

Posttraumatic nerve repair and prevention of muscle atrophy represent a major challenge of restorative medicine. Among the possible explanations for the incomplete restoration of very long term denervated muscle are failures of regenerating nerves to reach all of the atrophic muscle fibers and establish mature muscle–nerve contacts (Fu and Gordon, 1995; Irintchev, Draguhn, & Wernig, 1990). It was shown earlier that the sites of former endplates could be detected in surviving skeletal muscle fibers even after 17 months (Sunderland and Ray, 1950) and 2 years of muscle denervation (Debkov, Kostrominova, Borisov, & Carlson, 2001). Nevertheless, it is very difficult to assess the functional condition of those long-term denervated sites of former neuromuscular junctions with respect to their capacity to accept growing axons if reinnervation was to occur. Considerable interest exists in the potential therapeutic value of laser phototherapy for restoring denervated muscle resulting from peripheral nerve injury. We have previously suggested that the function of denervated muscles can be partially preserved by temporary prevention of denervation-induced biochemical changes (Rochkind et al., 2009). The function of denervated muscles can be restored, not completely but to a very substantial degree, by laser treatment, initiated at the earliest possible stage postinjury. Iyomasa, Garavelo, Iyomasa, Watanabe, and Issa (2009) found that the treatment of lesioned muscles with low-level He–Ne laser therapy could increase mitochondrial activity in muscular fibers and activate fibroblasts and macrophages and stimulate angiogenesis. Schwartz, Brodie, Appel, Kazimirsky, and Shainberg (2002) found that, after irradiation of muscle cell cultures (632 nm, 3 J/cm2, 20 mW), there was a rise in the levels of nerve growth factor, which is a neurotrophic factor secreted by skeletal muscles that influences the survival and regeneration of sympathetic and sensitive neurons in the peripheral nervous system. Other neurotropic growth factors are also biostimulated by laser therapy, such as GAP-43 (Shin et al., 2003) and fibroblast growth factor (Ihsan, 2005).

The data collected from different experimental studies support our results and help to understand the mechanism of influence of low-power laser irradiation (visible and near-infrared wavelengths) and muscle tissue. It has been demonstrated that reactive oxygen species (ROS) formation and augmented collagen synthesis are elicited by traumatic muscular injury, effects that were significantly decreased by laser treatment (Silveira et al., 2013). Evaluation of mitochondrial respiratory chain complexes and succinate dehydrogenase activities after traumatic muscular injury shows that the laser treatment may induce an increase in ATP synthesis, and that this may accelerate the muscle healing process (Silveira et al., 2009) and delay fusion of cultured myoblasts (Wollman and Rochkind, 1993). The increase in muscle fibers area and mitochondrial density after laser treatment was reported after muscle toxic injury (Amaral, Parizotto, & Salvini, 2001). The process of regeneration in denervated muscles was markedly enhanced in muscle that was irradiated by laser prior to injury, probably by the activation (stimulation of proliferation and/or differentiation) of cells in the muscles that are "recruited" and participate in the process of regeneration (Bibikova and Oron, 1995). In model of prolonged muscle ischemia, laser treatment decreased posttraumatic changes in CK and lactate dehydrogenase (Lakyová, Toporcer, Tomečková, Sabo, & Radoňak, 2010). A positive effect on muscle metabolism was found after cryolesion injury, whereas Cyclo-oxyge 2

(COX-2) immunoexpression was lower in laser-treated group. COX-2 is a key enzyme in conversion of arachidonic acid to prostanoids (Rennó et al., 2011). The expression of COX-2 is relevant to many pathological processes, including inflammation and tissue repair. In the present study, laser irradiation significantly increased CK activity and AChR level in time periods of 30–60 days in comparison with the nonirradiated gastrocnemius muscles. At the cell level, we also found increased DNA synthesis and CK activity in young and mature skeletal muscle. The induced biochemical changes may be attributed to trophic signal for increased activity of CK, thus preserving a reservoir of high-energy phosphate available for quick resynthesis of ATP. These findings are supported by early results by Bolognani and Volpi (1991) and Passarella et al. (1984) who showed that laser irradiation increased ATP production in the mitochondria. The present study and our previous publication (Rochkind et al., 2009) suggest that laser phototherapy may enhance biochemical activity of the muscle to overcome stress conditions.

REFERENCES

Almon, R. R., Andrew, C. G., & Appel, S. H. (1974). Acetylcholine receptor in normal and denervated slow and fast muscle. *Biochemistry, 13*, 5522–5528.

Amaral, A. C., Parizotto, N. A., & Salvini, T. F. (2001). Dose-dependency of low-energy HeNe laser effect in regeneration of skeletal muscle in mice. *Lasers in Medical Science, 16*, 44–51.

Battiston, B., Geuna, S., Ferrero, M., & Tos, P. (2005). Nerve repair by means of tubulization: Literature review and personal clinical experience comparing biological and synthetic conduits for sensory nerve repair. *Microsurgery, 25*, 258–267.

Bibikova, A., & Oron, U. (1995). Regeneration in denervated toad (Bufo viridis) gastrocnemius muscle and the promotion of the process by low energy laser irradiation. *The Anatomical Record, 241*, 123–128.

Bolognani, L., & Volpi, N. (1991). Low power laser enzymology: Reactivation of myosin ATPase by GaAs and HeNe lasers. In S. Passarella & E. Quadliariello (Eds.), *Basic and applied research in photobiology and photomedicine* (pp. 21–42). Italy: Bari.

Campbell, W. W. (2008). Evaluation and management of peripheral nerve injury. *Clinical Neurophysiology, 119*, 1951–1965.

Chin, H., & Almon, R. R. (1980). Fiber-type effects of castration on the cholinergic receptor population in skeletal muscle. *The Journal of Pharmacology and Experimental Therapeutics, 212*, 553–555.

Debkov, E. I., Kostrominova, T. Y., Borisov, A. B., & Carlson, B. M. (2001). Reparative myogenesis in long-term denervated skeletal muscles of adult rats results in a reduction of the satellite cell population. *The Anatomical Record, 263*, 139–154.

Fu, S. Y., & Gordon, T. (1995). Contributing factors to poor functional recovery after delayed nerve repair: Prolonged denervation. *The Journal of Neuroscience, 15*, 3886–3895.

Gigo-Benato, D., Geuna, S., & Rochkind, S. (2005). Phototherapy for enhancing peripheral nerve repair: A review of the literature. *Muscle & Nerve, 31*, 694–701.

Goldspink, D. F. (1976). The effects of denervation on protein turnover of rat skeletal muscle. *Journal of Biochemistry*, *156*, 71–80.

Guzzini, M., Raffa, S., Geuna, S., Nicolino, S., Torrisi, M. R., Tos, P., et al. (2008). Denervation-related changes in acetylcholine receptor density and distribution in the rat flexor digitorum sublimis muscle. *Italian Journal of Anatomy and Embryology*, *4*, 209–216.

Ihsan, F. R. M. (2005). Low-level laser therapy accelerates collateral circulation and enhances microcirculation. *Photomedicine and Laser Surgery*, *23*, 289–294.

Irintchev, A., Draguhn, A., & Wernig, A. (1990). Reinnervation and recovery of mouse soleus muscle after long-term denervation. *Neuroscience*, *39*, 231–243.

Iyomasa, D. M., Garavelo, I., Iyomasa, M. M., Watanabe, I., & Issa, J. P. M. (2009). Ultrastructural analysis of the low level laser therapy effects on the lesioned anterior tibial muscle in the gerbil. *Micron*, *40*, 413–418.

Kloosterboer, H., Faassen, H., Stroker-de Vries, S. A., & Hommes, F. A. (1979). Effect of cyclic nucleotides on weight of gastrocnemius and creatine kinase activity after denervation of muscle in young rats. *Biology of the Neonate*, *36*, 160–167.

Lakyová, L., Toporcer, T., Tomečková, V., Sabo, J., & Radoňak, J. (2010). Low-level laser therapy for protection against skeletal muscle damage after ischemia-reperfusion injury in rat hindlimbs. *Lasers in Surgery and Medicine*, *42*, 665–672.

Oliver, I. T. (1955). A spectrophotometric method for the determination of creatine phosphokinase and myokinase. *The Biochemical Journal*, *61*, 116–174.

Ontell, M. (1975). Evidence for myoblastic potential of satellite cells in denervated muscle. *Cell and Tissue Research*, *160*, 345–353.

Passarella, S., Casamassima, E., Molinari, S., Pastore, D., Quagliariello, E., Catalano, I. M., et al. (1984). Increase of proton electrochemical potential and ATP synthesis in rat liver mitochondria irradiated in vitro by helium-neon laser. *FEBS Letters*, *175*, 95–99.

Rennó, A. C., Toma, R. L., Feitosa, S. M., Fernandes, K., Bossini, P. S., de Oliveira, P., et al. (2011). Comparative effects of low-intensity pulsed ultrasound and low-level laser therapy on injured skeletal muscle. *Photomedicine and Laser Surgery*, *29*, 5–10.

Rochkind, S., Geuna, S., & Shainberg, A. (2009). The role of phototherapy in peripheral nerve regeneration and muscle preservation. *International Review of Neurobiology*, *87*, 445–464.

Rosaki, S. B. (1967). Creatine phosphokinase—A spectrophotometric method with improved sensitivity. *American Journal of Clinical Pathology*, *50*, 89–97.

Schwartz, F., Brodie, C., Appel, E., Kazimirsky, G., & Shainberg, A. (2002). Effect of helium/neon laser irradiation on nerve growth factor synthesis and secretion in skeletal muscle cultures. *The Journal of Photochemistry and Photobiology B*, *66*, 195–200.

Shainberg, A., & Burstein, M. (1976). Decrease of acetylcholine receptor synthesis in muscle cultures by electrical stimulation. *Nature*, *264*, 368–369.

Shainberg, A., Yagil, G., & Yaffe, D. (1971). Aterations of enzymatic activities during muscle differentiation in vitro. *Developmental Biology*, *25*, 1–29.

Shin, D. H., Lee, E., Hyun, J., Lee, S. J., Chang, Y. P., Kim, J., et al. (2003). Growth associated protein-43 is elevated in the injured rat sciatic nerve after low power irradiation. *Neuroscience Letters*, *344*, 71–74.

Silveira, P. C., da Silva, L. A., Pinho, C. A., De Souza, P. S., Ronsani, M. M., da Luz Scheffer, D., et al. (2013). Effects of low-level laser therapy (GaAs) in an animal model of muscular damage induced by trauma. *Lasers in Medical Science*, *28*, 431–436.

Silveira, P. C., Silva, L. A., Fraga, D. B., Freitas, T. P., Streck, E. L., & Pinho, R. (2009). Evaluation of mitochondrial respiratory chain activity in muscle healing by low-level laser therapy. *The Journal of Photochemistry and Photobiology B*, *95*, 89–92.

Sunderland, S., & Ray, L. J. (1950). Denervation changes in mammalian striated muscle. *Journal of Neurology, Neurosurgery, and Psychiatry, 13*, 159–177.

Wada, K., Katsuta, S., & Soya, H. (2008). Formation process and fate of the nuclear chain after injury in regenerated myofiber. *The Anatomical Record, 291*, 122–128.

Wollman, Y., & Rochkind, S. (1993). Muscle fiber formation in vitro is delayed by low power laser irradiation. *The Journal of Photochemistry and Photobiology B, 17*, 287–290.

Yaffe, D. (1969). Cellular aspects of muscle differentiation in vitro. *Current Topics in Developmental Biology, 4*, 37–77.

CHAPTER FIVE

Electrical Stimulation for Promoting Peripheral Nerve Regeneration

Kirsten Haastert-Talini[*,†,1], Claudia Grothe[*,†]
[*]Hannover Medical School, Institute of Neuroanatomy, Hannover, Germany
[†]Center for Systems Neuroscience (ZSN), Hannover, Germany
[1]Corresponding author: e-mail address: Haastert.kirsten@mh-hannover.de

Contents

Abstract

The peripheral nervous system has the intrinsic capacity to regenerate axons into target tissues, and peripheral nerves severely damaged or transected can be reconstructed by microsurgical techniques. The aim of peripheral nerve surgery is to pave way for fast and most possible thorough functional recovery. However, full functional recovery is rarely seen and several reasons for this have already been discovered. Based on these discoveries, therapeutic strategies supplementary to nerve microsurgery have been conceived with electrical stimulation of the denervated muscles or the proximal nerve stump or reconstructed area itself being among them. This chapter shortly describes the commonly accepted reasons for incomplete functional recovery and reviews the effects of varying electrical stimulation paradigms on the essentials for axonal regeneration and functional target reinnervation. We conclude the chapter with promising examples where electrical stimulation did already demonstrate to accelerate and increase functional recovery in the clinic.

International Review of Neurobiology, Volume 109
ISSN 0074-7742
http://dx.doi.org/10.1016/B978-0-12-420045-6.00005-5

1. INTRODUCTION

The degree of functional recovery after peripheral nerve reconstruction is often nonsatisfying although the neurons of the peripheral nervous system have the machinery to regenerate their axons and no general inhibitory environment develops in the distal nerve stumps or target tissue (Deumens et al., 2010). Functional recovery is most complete if regeneration occurs after crush injury, and it can still reach a valuable degree after tension-free end-to-end suture (Deumens et al., 2010). But whenever severe trauma or tumor resection leads to long distances between transected peripheral nerve stumps, the mechanisms impeding timely and appropriate target reinnervation become seriously obstructing (Ray & Mackinnon, 2010).

As reasons for insufficient functional recovery, the increasing irreversible muscle atrophy with time and the long period axonal sprouts, extending from far proximal lesion sites, need to reach distal motor targets were previously discussed (Fu & Gordon, 1997; Gordon, Sulaiman, & Ladak, 2009). Recently, *Tessa Gordon* and colleagues have defined the main factors impeding axonal regeneration in more detail as (1) the slow growth of regenerating axons across surgical coaptation sites and nerve gaps, (2) the decline in regenerative capacity of axotomized neurons with time, and (3) the limited time in which denervated Schwann cells can successfully contribute a regenerative milieu for regenerating axons (Gordon, Sulaiman, & Ladak, 2009; Gordon, Tyreman, & Raji, 2011).

Furthermore, only when regenerating axons are not misguided into the wrong distal pathways and appropriate target reinnervation occurs, a sufficient degree of functional recovery can be expected (Valero-Cabre & Navarro, 2002).

Electrical stimulation was conceived as one therapeutic strategy to expedite regenerative axonal growth of regenerating axons across surgical repair sites and to improve regeneration accuracy especially of motor axons (Gordon, Sulaiman, & Ladak, 2009).

Different ways to apply electrical stimulation protocols with the aim to increase peripheral nerve regeneration are described in the following.

2. WAYS TO APPLY ELECTRICAL STIMULI TO THE RECONSTRUCTED NERVE

2.1. Transcutaneous/percutaneous stimulation

Transcutaneous electrical stimulation (TENS) is used clinically as rehabilitation treatment in order to recover impaired muscle functionality

(Gigo-Benato et al., 2010). However, TENS was shown to impair axonal regeneration after crush lesion (Baptista et al., 2008). This result was confirmed by a second study demonstrating that TENS to the tibialis anterior muscle first administered 3 days after sciatic nerve crush and then every second day for a 14-day-period inhibited functional recovery of the neuromuscular system (Gigo-Benato et al., 2010).

That instead continuous percutaneous stimulation with a low intensity could be advantageous for nerve regeneration has been reported also from a crush injury model (Mendonca, Barbieri, & Mazzer, 2003).

On the opposite, percutaneous electrical stimulation 15 min/day starting 1 week after nerve repair did reduce the macroscopic regeneration across 10-mm nerve gaps bridged with silicone tubes. However, within the nerves that successfully bridged the gap, the morphometrical parameters for nerve regeneration were significantly increased in comparison to nonstimulated nerves (Chen et al., 2001).

Comparison of different percutaneous stimulation protocols of pulse current versus direct current stimulation revealed that pulse stimulation increased nerve fiber density and direct current stimulation decreased nerve fiber density (Cheng & Lin, 2004). It has further been demonstrated that percutaneous stimulation can support axonal regeneration and maturation only when stimulation protocols are well controlled and stimulation is restricted to minimal repetitions and minimal intensities, otherwise the treatment could be detrimental (Lu et al., 2009). Recently, a percutaneous electrical stimulation protocol (15 min every second day) has been described to improve peripheral nerve regeneration across silicone tube-bridged nerve gaps (10 mm) in a diabetic rat model (Yao et al., 2012). However, a single, brief and more controlled stimulation protocol directly applied to the reconstructed nerve at the time of peripheral nerve reconstruction has been developed over the past decade and is described in the following.

2.2. Direct, brief, low-frequency electrical stimulation

Direct low-frequency (20 Hz) electrical stimulation of the proximal nerve stump for 1 h (two to threefold threshold current) together with nerve reconstruction surgery has been established as the standard procedure (Gordon, Sulaiman, & Ladak, 2009). It has been evidenced that the treatment accelerates axonal outgrowth across the suture sites after end-to-end coaptation or small gap reconstruction and that functional motor recovery is supported as well (Ahlborn, Schachner, & Irintchev, 2007; Al-Majed, Neumann, Brushart, & Gordon, 2000; Brushart et al., 2002; Gordon, Sulaiman, & Ladak, 2009; Vivo et al., 2008). Furthermore, a regeneration-promoting effect after 10 mm autotransplantation in the rat femoral nerve model has been described (Huang et al., 2009).

Highlighting the importance of a defined stimulation protocol, low-frequency stimulation for 1 h has been demonstrated to be as effective as continuous stimulation over 2 weeks for motoneurons (Al-Majed, Neumann, et al., 2000), but for sensory neurons, the positive effect of accelerated axonal regeneration across a nerve repair site is abolished when the stimulation period is extended over 1 h (Geremia, Gordon, Brushart, Al-Majed, & Verge, 2007). Furthermore, chronic electrical stimulation (1 h daily) was described to be less effective than brief acute electrical stimulation with regard to muscle reinnervation and axonal regeneration (Asensio-Pinilla, Udina, Jaramillo, & Navarro, 2009).

The regeneration-promoting effects of the acute direct electrical stimulation paradigm has also been successfully tested in animal models for long-gap repair. Our own studies clearly demonstrated that the brief, low-frequency direct stimulation protocol is sufficient to accelerate long-distance regeneration of motor and sensory axons resulting in an increase of functional regeneration especially after nerve autotransplantation (13-mm gap lengths) in rats (Haastert-Talini et al., 2011).

But also for 15-mm rat sciatic nerve gap reconstruction with longitudinally oriented microchannels, it has been shown that axonal and motor regeneration rates after electrical stimulation reach levels comparable to the gold standard nerve autotransplantation without stimulation (Huang, Lu, et al., 2010).

Tissue engineering of peripheral nerves is an attempt of high actuality (Grothe et al., 2012) and electrical stimulation may provide a valuable cotreatment to it.

2.3. Electrical stimulation via the synthetic nerve graft

Besides the transcutaneous/percutaneous or direct application of electrical stimuli to the reconstructed nerve also induction of an electrical field along the nerve gaps bridged with synthetic tubes has been already tested decades ago. Electric fields along nerve guidance tubes can be elicited by the use of piezoelectric tube material. This material generates transient electrical charges upon deformation (Aebischer, Valentini, Dario, Domenici, Guenard, et al., 1987). Bridging 4-mm nerve gaps in the mouse with electrically poled piezoelectric tubes significantly increased regeneration of myelinated axons in comparison to bridging the defects with electrically inert unpoled tubes made of the same material (Aebischer Valentini, Dario, Domenici, & Galletti., 1987).

Another way to induce an electrical field along the nerve bridge is to insert a cathode electrode at the midpoint of a silicone tube and the anode electrode distal to it creating a constant negative electric field in distal direction. Using the described technique, two independent studies were performed which provided the electrical stimulation over 21 days in a 5-mm nerve gap and demonstrated opposing results. The first study described a tremendous increase of the number of regenerated axons proximal to the cathode (Roman, Strahlendorf, Coates, & Rowley, 1987) and the other deleterious effects on the axonal regeneration distal to the cathode (Hanson & McGinnis, 1994). It had to be concluded that this way to apply electrical stimulation to a regenerating nerve is difficult to control.

Current developments to apply electrical stimuli via conductive nerve guidances channels and scaffolds have been recently discussed in a comprehensive review (Ghasemi-Mobarakeh et al., 2011). This field of bioengineering is still evolving, and further investigations are needed to design the optimal scaffold combining intrinsic electrical stimulation paradigms for optimal support of regeneration and axonal guidance.

Very recently, however, it has been described that copper loop-electrodes secured at the proximal and distal end of a conductive microchannel scaffold could be used for brief (1 h), low-frequency (20 Hz) electrical stimulation to induce an electrical field along a 15-mm rat sciatic nerve gap (Huang et al., 2012). Repeated electrical stimulation (every second day at eight times) together with nerve reconstruction by means of the microstructured conductive nerve guide demonstrated high efficacy to promote axonal and functional recovery of large nerve defects (Huang et al., 2012).

The bioengineering field provides diverse possibilities to combine regeneration-supportive strategies in an optimized nerve reconstruction treatment, and further developments are needed. However, electrical stimulation has already been combined to other treatments that showed regeneration-promoting activities as well.

2.4. Electrical stimulation combined with other treatments

We performed a study in which we combined the brief, low-frequency direct stimulation protocol with the transplantation of silicone tubes containing naïve Schwann cells or Schwann cells genetically modified to over-express fibroblast growth factor-2 (Haastert-Talini et al., 2011). The latter treatment had demonstrated regeneration-promoting potential before (Haastert, Lipokatic, Fischer, Timmer, & Grothe, 2006; Haastert, Ying, Grothe, & Gomez-Pinilla, 2008). Electrical stimulation plus transplantation of naïve Schwann cells increased the regeneration of gap-bridging nerve tissue in comparison to nerve reconstruction with cell-free tubes. Unexpectedly, no further increase of macroscopic tissue regeneration after transplantation of FGF-2 overexpressing Schwann cells was found. However, electrical stimulation combined with the chosen gene therapy paradigm resulted in a high rate of regenerated nerves that functionally reconnected to the target muscle (Haastert-Talini et al., 2011).

Combination of electrical stimulation with a gene therapy approach was also tested by another group. After sciatic nerve crush injury, either electrical stimulation or adenoviral brain-derived neurotrophic factor (BDNF) gene therapy was applied and compared to a combined treatment (Alrashdan et al., 2011). In comparison to no treatment, either electrical stimulation or BDNF supplementation alone resulted in an increase of axonal regeneration but the combination of both eliminated the positive effects (Alrashdan et al., 2011). The study again revealed that combinatory treatments have to be carefully adjusted to each other to allow synergistic effects.

Moderate-intensity treadmill training as another treatment shown to increase axonal and functional recovery has also been combined with acute brief electrical stimulation in a model of sciatic nerve transection and suture (Asensio-Pinilla et al., 2009). Those treatment strategies synergistically supported the regeneration processes by shortening the period of staggered axonal outgrowth and maintaining regeneration-supportive stimuli mediated from the target tissue (Asensio-Pinilla et al., 2009). While in the initial study, the accelerated recovery of nociception and muscle reinnervation was recognized as a positive outcome of the study (Asensio-Pinilla et al., 2009), especially the effect of the combined treatment on the development of neuropathic pain was investigated in a recent study (Cobianchi, Casals-Diaz, Jaramillo, & Navarro, 2013). The development of neuropathic pain is a sign of maladaptation of nociceptors during regeneration-related plastic changes and can be modulated if excessive sprouting of sensory fibers is limited (Cobianchi et al., 2013). Acute electrical stimulation followed by high-intensity short-term treadmill training has been shown to be effective to avoid development of neuropathic pain (Cobianchi et al., 2013). The regeneration-promoting effects of both treatments are neurotrophin dependent and the signal cascades seem to be balanced to provide synergistic support of functional recovery (Cobianchi et al., 2013).

The mechanisms underlying the regeneration-promoting effect of electrical stimulation are subject of the following section.

3. BIOLOGICAL EFFECTS OF ELECTRICAL STIMULATION OF INJURED PERIPHERAL NERVES

3.1. General effect on axonal regeneration

The period of staggered axonal regeneration is significantly shortened (Al-Majed, Neumann, et al., 2000), but not the speed of axonal elongation (Brushart et al., 2002). Furthermore, the correct reinnervation of axonal motor and sensory pathways is supported by brief electrical stimulation (Al-Majed, Neumann, et al., 2000; Brushart, Jari, Verge, Rohde, & Gordon, 2005). The positive effect on correct muscle reinnervation can be further triggered with muscle-derived signals initiated by treadmill exercise (Asensio-Pinilla et al., 2009).

The electrical treatment is accelerating the number of axons crossing the nerve repair site as a function of time, but it is not accelerating axonal transport mechanisms or the speed of axonal elongation within the distal nerve stump (Gordon, Brushart, & Chan, 2008).

Electrical stimulation applied in nerve gap repair also accelerates regrowth of nerve tissue into nerve guidance channels and anticipates maturation and myelination of regenerated axons (Haastert-Talini et al., 2011; Singh et al., 2012). When combined with silicone tube repair of short nerve gaps in mice (Singh et al., 2012) or lengthy nerve gaps with autotransplants in rats (Haastert-Talini et al., 2011), electrical stimulation does not only improve motor recovery but also anticipates sensory recovery. Furthermore, as mentioned above, earlier reinnervation of sensory targets by electrical stimulation has recently been linked to reduced development of neuropathic pain after peripheral nerve injury (Cobianchi et al., 2013).

Incomplete functional recovery is not only attributed to impeded axonal regeneration but also to the misdirection of axons into inappropriate motor targets, for example, antagonist muscles (Valero-Cabre & Navarro, 2002). Whether electrical stimulation can positively influence the reinnervation of appropriate motor targets was addressed in a model of selective prestimulation of a single sciatic nerve branch prior to transection of the complete nerve trunk at mid-thigh level (Hamilton et al., 2011). Interestingly, electrical prestimulation did not reduce but increase misdirection of regenerating motor axons into inappropriate motor targets (Hamilton et al., 2011). This finding not only strengthens again the outgrowth-promoting effect of electrical stimulation but also demonstrates that, after reinnervation occurred, no overall functional benefit occurs from the treatment (Gordon, Amirjani, Edwards, & Chan, 2010; Hamilton et al., 2011). However, shortening of regeneration times is of high value for patients suffering from peripheral nerve lesions.

3.2. Impact of electrical stimulation on expression of neurotrophic factors

It is well demonstrated that short-term low-frequency electrical stimulation proximal to the suture site of freshly transected and readapted nerves increases the expression of BDNF and its specific receptor trkB in axotomized motoneurons (Al-Majed, Brushart, & Gordon, 2000). The upregulation of both BDNF and trkB is crucially involved in the upregulation of regeneration-associated genes such as Tα1-tubulin and growth-associated protein 43 (GAP-43) as well as the subsequent faster axonal elongation (Al-Majed, Tam, & Gordon, 2004; Gordon, Sulaiman, & Ladak, 2009).

With sophisticated studies, it has been proven that (i) the upregulation of neurotrophic factors is essential to mediate the cellular mechanisms of accelerated axonal elongation and target reinnervation after electrical stimulation (Gordon, 2010) and that (ii) the cellular mechanism is mainly based on the activation of trkB signaling with ligands produced either by Schwann cells or the regenerating axons themselves (English, Schwartz, Meador, Sabatier, & Mulligan, 2007).

It has been discovered that prior to the upregulation of BDNF and the trkB receptor, electrical stimulation induces an increased influx of calcium into the neurons (Gordon, 2009; Wenjin et al., 2011). The calcium influx induces phosphorylation of the Ca^{2+}-activated extracellular signal regulated kinase, Erk, which is another prerequisite for the elevated BDNF expression after electrical stimulation (Wenjin et al., 2011). Furthermore, the described calcium influx is followed by an increase of intracellular cAMP levels, and downstream of cAMP, protein kinase A promotes expression of the regeneration-associated genes for axonal elongation (Gordon, 2009).

In our own investigation of the effect of electrical stimulation after long (13 mm)-gap nerve tube repair, no relative upregulation of BDNF–mRNA levels related to electrical stimulation in sensory or motoneurons was seen but a prolonged BDNF expression which was accompanied by a significant upregulation of GAP-43 mRNA levels within the lumbar spinal cord 2 weeks after surgery and stimulation (Haastert-Talini et al., 2011).

Recently, it has been demonstrated that the neuronal expression of not only BDNF but also additional neurotrophic factors is increased after electrical stimulation *in vivo*. In the ventral horn Neurotrophin-3 and in the dorsal root ganglia nerve growth factor (NGF) and glia-derived neurotrophic factor showed significant upregulation after acute electrical stimulation (Cobianchi et al., 2013).

The optimal orchestration of the neurotrophin-dependent signals not only in neurons but also in Schwann cells finally leads to successful nerve regeneration.

3.3. Impact of electrical stimulation on Schwann cells

Electrical stimulation, when applied directly to the nerve at levels proximal to the nerve suture, is sufficient to increase neurotrophic factor levels in the distal nerve as well (Wang et al., 2009). Partly, the increased amount of neurotrophic factors is of neural origin and subsequently stimulates the biological activity of Schwann cells in the distal nerve (Wang et al., 2009).

As downstream effector of neurotrophin signaling in Schwann cells of motor pathways, the HNK-1 carbohydrate has been discussed. Its expression has been shown to contribute to the preferential motor reinnervation (PMR) phenomenon in the femoral nerve model, and this specific expression pattern is significantly enhanced by short-term electrical stimulation of acutely injured and reconnected nerves (Eberhardt et al., 2006). Again, the importance of BDNF signaling for the translation of electrical stimulation into increased axonal regeneration has been evidenced by the abolishment of increased HNK-1 expression after electrical stimulation in the distal nerve stump in trkB-deficient mice (Eberhardt et al., 2006).

With regard to PMR, it is of high importance to recognize that the guiding Schwann cells and regenerating axons can influence each other reciprocally. Polysialic acid (PSA) linked to the neural cell-adhesion molecule (NCAM) has been shown to crucially contribute to the PMR phenomenon (Franz, Rutishauser, & Rafuse, 2005). Subsets of femoral motoneurons, which regenerate preferentially into the motor branch instead of the sensory branch of the femoral nerve, show an upregulation of PSA–NCAM (Franz et al., 2005). Electrical stimulation increases the upregulation of PSA and thereby further improves preferential motor axonal regeneration into motor target tissue (Franz, Rutishauser, & Rafuse, 2008). However, it has to be remembered that if appropriate and inappropriate motor targets are located distal to the lesioned nerve, electrical stimulation does not positively influence appropriate reinnervation of original motor targets (Hamilton et al., 2011).

BDNF-secretion from regenerating axons after electrical stimulation also advances the myelination of the regrown axons (Wan, Xia, & Ding, 2010). This has been demonstrated *in vitro* and *in vivo* where increased BDNF levels after electrical stimulation were identified to accelerate myelination by advancing Schwann cell polarization and initiation of myelination resulting in a higher degree of myelination at earlier times points as in nonstimulated controls (Wan et al., 2010).

Electrical stimulation also has a high potential to directly influence Schwann cell behavior. *In vitro* electrical stimulation resulted in increased proliferation of Schwann cells cultured on conductive membranes (Huang, Hu, et al., 2010). Furthermore, the expression of NGF and BDNF was significantly increased in Schwann cells due to electrical stimulation as was the secretion of both factors over a period of 36 h after stimulation (Huang, Hu, et al., 2010). That application of electric fields of varying intensities and frequencies to cultured neonatal rat Schwann cells significantly increased their production and release of NGF in an intensity- and frequency-dependent manner was also shown in a second *in vitro* study (Huang, Ye, Hu, Lu, & Luo, 2010). Again, as also described for neuronal cells before, an increase in Schwann cell intracellular calcium levels related to electrical stimulation demonstrated to be involved in elevated expression and exocytosis levels of NGF (Huang, Ye, et al., 2010).

That the effect of electrical stimulation on Schwann cells has a regeneration-promoting character, additional to its stimulation of neurons, has been demonstrated with another *in vitro* study (Koppes, Seggio, & Thompson, 2011). Neurite outgrowth from sensory dorsal root ganglion neurons was significantly stronger when electrical stimulation was applied to neurons cocultured with Schwann cells in comparison to stimulated neurons cultured alone or unstimulated neurons in cocultures with Schwann cells (Koppes et al., 2011).

Over the past decade, plenty knowledge has been achieved on the effects of electrical stimulation which qualified the treatment as promising cotherapy for the clinic.

4. CLINICAL EXPERIENCES

Acute brief low-frequency electrical stimulation has also been tested as a clinical tool already. Directly applied after open carpal tunnel release surgery, the treatment has been demonstrated to be well tolerated by the patients (Gordon, Brushart, Amirjani, & Chan, 2007).

In a randomized pilot study, patients who suffered from substance loss in their median nerves due to its severe compression in the carpal tunnel were included and electrical stimulation was applied to the compression-released median nerve (Gordon, Chan, et al., 2009; Gordon et al., 2010). This treatment significantly increased the number of functional motor units in the thenar muscle and resulted in significantly accelerated reinnervation (several months) of the distal targets in comparison to nonstimulated patients (Gordon, Chan, et al., 2009; Gordon et al., 2010).

These positive clinical reports together with increasing knowledge on the regeneration-promoting mechanisms initiated by electrical stimulation of the peripheral nerve will probably support a wider use of electrical stimulation as cotreatment for nerve reconstruction surgery in the future.

ACKNOWLEDGMENT

Funded by the European Community's Seventh Framework Programme (FP7-HEALTH-2011) under grant agreement n°278612.

Conflict of interest disclosure: No conflicts of interests exist for any of the authors.

REFERENCES

Aebischer, P., Valentini, R. F., Dario, P., Domenici, C., & Galletti, P. M. (1987). Piezoelectric guidance channels enhance regeneration in the mouse sciatic nerve after axotomy. *Brain Research, 436*(1), 165–168, 0006-8993(87)91570-8 [pii].

Aebischer, P., Valentini, R. F., Dario, P., Domenici, C., Guenard, V., Winn, S. R., et al. (1987). Piezoelectric nerve guidance channels enhance peripheral nerve regeneration. *ASAIO Transactions, 33*(3), 456–458.

Ahlborn, P., Schachner, M., & Irintchev, A. (2007). One hour electrical stimulation accelerates functional recovery after femoral nerve repair. *Experimental Neurology, 208*(1), 137–144.

Al-Majed, A. A., Brushart, T. M., & Gordon, T. (2000). Electrical stimulation accelerates and increases expression of BDNF and trkB mRNA in regenerating rat femoral motoneurons. *The European Journal of Neuroscience, 12*(12), 4381–4390.

Al-Majed, A. A., Neumann, C. M., Brushart, T. M., & Gordon, T. (2000). Brief electrical stimulation promotes the speed and accuracy of motor axonal regeneration. *The Journal of Neuroscience, 20*(7), 2602–2608.

Al-Majed, A. A., Tam, S. L., & Gordon, T. (2004). Electrical stimulation accelerates and enhances expression of regeneration-associated genes in regenerating rat femoral motoneurons. *Cellular and Molecular Neurobiology, 24*(3), 379–402.

Alrashdan, M. S., Sung, M. A., Kim Kwon, Y., Chung, H. J., Kim, S. J., & Lee, J. H. (2011). Effects of combining electrical stimulation with BDNF gene transfer on the regeneration of crushed rat sciatic nerve. *Acta Neurochirurgica, 153*(10), 2021–2029. http://dx.doi.org/10.1007/s00701-011-1054-x.

Asensio-Pinilla, E., Udina, E., Jaramillo, J., & Navarro, X. (2009). Electrical stimulation combined with exercise increase axonal regeneration after peripheral nerve injury. *Experimental Neurology, 219*(1), 258–265.

Baptista, A. F., Gomes, J. R., Oliveira, J. T., Santos, S. M., Vannier-Santos, M. A., & Martinez, A. M. (2008). High- and low-frequency transcutaneous electrical nerve stimulation delay sciatic nerve regeneration after crush lesion in the mouse. *Journal of the Peripheral Nervous System, 13*(1), 71–80.

Brushart, T. M., Hoffman, P. N., Royall, R. M., Murinson, B. B., Witzel, C., & Gordon, T. (2002). Electrical stimulation promotes motoneuron regeneration without increasing its speed or conditioning the neuron. *The Journal of Neuroscience, 22*(15), 6631–6638.

Brushart, T. M., Jari, R., Verge, V., Rohde, C., & Gordon, T. (2005). Electrical stimulation restores the specificity of sensory axon regeneration. *Experimental Neurology, 194*(1), 221–229.

Chen, Y. S., Hu, C. L., Hsieh, C. L., Lin, J. G., Tsai, C. C., Chen, T. H., et al. (2001). Effects of percutaneous electrical stimulation on peripheral nerve regeneration using silicone rubber chambers. *Journal of Biomedical Materials Research, 57*(4), 541–549.

Cheng, W. L., & Lin, C. C. (2004). The effects of different electrical stimulation protocols on nerve regeneration through silicone conduits. *Journal of Trauma, 56*(6), 1241–1246, 00005373-200406000-00013 [pii].

Cobianchi, S., Casals-Diaz, L., Jaramillo, J., & Navarro, X. (2013). Differential effects of activity dependent treatments on axonal regeneration and neuropathic pain after peripheral nerve injury. *Experimental Neurology, 240*, 157–167. http://dx.doi.org/10.1016/j.expneurol.2012.11.023, S0014-4886(12)00440-2 [pii].

Deumens, R., Bozkurt, A., Meek, M. F., Marcus, M. A., Joosten, E. A., Weis, J., et al. (2010). Repairing injured peripheral nerves: Bridging the gap. *Progress in Neurobiology, 92*(3), 245–276. http://dx.doi.org/10.1016/j.pneurobio.2010.10.002, S0301-0082(10) 00172-3 [pii].

Eberhardt, K. A., Irintchev, A., Al-Majed, A. A., Simova, O., Brushart, T. M., Gordon, T., et al. (2006). BDNF/TrkB signaling regulates HNK-1 carbohydrate expression in regenerating motor nerves and promotes functional recovery after peripheral nerve repair. *Experimental Neurology, 198*(2), 500–510.

English, A. W., Schwartz, G., Meador, W., Sabatier, M. J., & Mulligan, A. (2007). Electrical stimulation promotes peripheral axon regeneration by enhanced neuronal neurotrophin signaling. *Developmental Neurobiology, 67*(2), 158–172.

Franz, C. K., Rutishauser, U., & Rafuse, V. F. (2005). Polysialylated neural cell adhesion molecule is necessary for selective targeting of regenerating motor neurons. *The Journal of Neuroscience, 25*(8), 2081–2091.

Franz, C. K., Rutishauser, U., & Rafuse, V. F. (2008). Intrinsic neuronal properties control selective targeting of regenerating motoneurons. *Brain, 131*(Pt 6), 1492–1505.

Fu, S. Y., & Gordon, T. (1997). The cellular and molecular basis of peripheral nerve regeneration. *Molecular Neurobiology, 14*(1–2), 67–116.

Geremia, N. M., Gordon, T., Brushart, T. M., Al-Majed, A. A., & Verge, V. M. (2007). Electrical stimulation promotes sensory neuron regeneration and growth-associated gene expression. *Experimental Neurology, 205*(2), 347–359.

Ghasemi-Mobarakeh, L., Prabhakaran, M. P., Morshed, M., Nasr-Esfahani, M. H., Baharvand, H., Kiani, S., et al. (2011). Application of conductive polymers, scaffolds and electrical stimulation for nerve tissue engineering. *Journal of Tissue Engineering and Regenerative Medicine, 5*(4), e17–e35. http://dx.doi.org/10.1002/term.383.

Gigo-Benato, D., Russo, T. L., Geuna, S., Domingues, N. R., Salvini, T. F., & Parizotto, N. A. (2010). Electrical stimulation impairs early functional recovery and accentuates skeletal muscle atrophy after sciatic nerve crush injury in rats. *Muscle & Nerve, 41*(5), 685–693. http://dx.doi.org/10.1002/mus.21549.

Gordon, T. (2009). The role of neurotrophic factors in nerve regeneration. *Neurosurgical Focus, 26*(2), E3.

Gordon, T. (2010). The physiology of neural injury and regeneration: The role of neurotrophic factors. *Journal of Communication Disorders, 43*(4), 265–273. http://dx.doi.org/10.1016/j.jcomdis.2010.04.003, S0021-9924(10)00021-3 [pii].

Gordon, T., Amirjani, N., Edwards, D. C., & Chan, K. M. (2010). Brief post-surgical electrical stimulation accelerates axon regeneration and muscle reinnervation without affecting the functional measures in carpal tunnel syndrome patients. *Experimental Neurology*, *223*(1), 192–202. http://dx.doi.org/10.1016/j.expneurol.2009.09.020, S0014-4886(09) 00407-5 [pii].

Gordon, T., Brushart, T. M., Amirjani, N., & Chan, K. M. (2007). The potential of electrical stimulation to promote functional recovery after peripheral nerve injury—Comparisons between rats and humans. *Acta Neurochirurgica Supplement*, *100*, 3–11.

Gordon, T., Brushart, T. M., & Chan, K. M. (2008). Augmenting nerve regeneration with electrical stimulation. *Neurological Research*, *30*(10), 1012–1022.

Gordon, T., Chan, K. M., Sulaiman, O. A., Udina, E., Amirjani, N., & Brushart, T. M. (2009). Accelerating axon growth to overcome limitations in functional recovery after peripheral nerve injury. *Neurosurgery*, *65*(4 Suppl.), A132–A144. http://dx.doi.org/ 10.1227/01.NEU.0000335650.09473, D3 00006123-200910001-00022 [pii].

Gordon, T., Sulaiman, O. A., & Ladak, A. (2009). Chapter 24: Electrical stimulation for improving nerve regeneration: Where do we stand? *International Review of Neurobiology*, *87*, 433–444.

Gordon, T., Tyreman, N., & Raji, M. A. (2011). The basis for diminished functional recovery after delayed peripheral nerve repair. *Journal of Neuroscience*, *31*(14), 5325–5334. http://dx.doi.org/10.1523/JNEUROSCI.6156-10.2011, 31/14/5325 [pii].

Grothe, C., Haastert-Talini, K., Freier, T., Navarro, X., Dahlin, L. B., Salgado, A., et al. (2012). BIOHYBRID—Biohybrid templates for peripheral nerve regeneration. *Journal of the Peripheral Nervous System*, *17*(2), 220–222. http://dx.doi.org/10.1111/j.1529-8027.2012.00399.x.

Haastert, K., Lipokatic, E., Fischer, M., Timmer, M., & Grothe, C. (2006). Differentially promoted peripheral nerve regeneration by grafted Schwann cells over-expressing different FGF-2 isoforms. *Neurobiology of Disease*, *21*(1), 138–153.

Haastert, K., Ying, Z., Grothe, C., & Gomez-Pinilla, F. (2008). The effects of FGF-2 gene therapy combined with voluntary exercise on axonal regeneration across peripheral nerve gaps. *Neuroscience Letters*, *443*(3), 179–183.

Haastert-Talini, K., Schmitte, R., Korte, N., Klode, D., Ratzka, A., & Grothe, C. (2011). Electrical stimulation accelerates axonal and functional peripheral nerve regeneration across long gaps. *Journal of Neurotrauma*, *28*(4), 661–674. http://dx.doi.org/10.1089/ neu.2010.1637.

Hamilton, S. K., Hinkle, M. L., Nicolini, J., Rambo, L. N., Rexwinkle, A. M., Rose, S. J., et al. (2011). Misdirection of regenerating axons and functional recovery following sciatic nerve injury in rats. *The Journal of Comparative Neurology*, *519*(1), 21–33. http://dx. doi.org/10.1002/cne.22446.

Hanson, S. M., & McGinnis, M. E. (1994). Regeneration of rat sciatic nerves in silicone tubes: Characterization of the response to low intensity d.c. stimulation. *Neuroscience*, *58*(2), 411–421.

Huang, J., Hu, X., Lu, L., Ye, Z., Wang, Y., & Luo, Z. (2009). Electrical stimulation accelerates motor functional recovery in autograft-repaired 10 mm femoral nerve gap in rats. *Journal of Neurotrauma*, *26*(10), 1805–1813. http://dx.doi.org/10.1089/neu.2008-0732.

Huang, J., Hu, X., Lu, L., Ye, Z., Zhang, Q., & Luo, Z. (2010). Electrical regulation of Schwann cells using conductive polypyrrole/chitosan polymers. *Journal of Biomedical Materials Research Part A*, *93*(1), 164–174. http://dx.doi.org/10.1002/jbm.a.32511.

Huang, J., Lu, L., Hu, X., Ye, Z., Peng, Y., Yan, X., et al. (2010). Electrical stimulation accelerates motor functional recovery in the rat model of 15-mm sciatic nerve gap bridged by scaffolds with longitudinally oriented microchannels. *Neurorehabilitation and Neural Repair*, *24*(8), 736–745. http://dx.doi.org/10.1177/1545968310368686, 1545968310368686 [pii].

Huang, J., Lu, L., Zhang, J., Hu, X., Zhang, Y., Liang, W., et al. (2012). Electrical stimulation to conductive scaffold promotes axonal regeneration and remyelination in a rat
model of large nerve defect. *PLoS One*, 7(6), e39526. http://dx.doi.org/10.1371/jour-
nal.pone.0039526, PONE-D-11-19433 [pii].

Huang, J., Ye, Z., Hu, X., Lu, L., & Luo, Z. (2010). Electrical stimulation induces calcium-
dependent release of NGF from cultured Schwann cells. *Glia*, 58(5), 622–631. http://dx.
doi.org/10.1002/glia.20951.

Koppes, A. N., Seggio, A. M., & Thompson, D. M. (2011). Neurite outgrowth is significantly increased by the simultaneous presentation of Schwann cells and moderate exog-
enous electric fields. *Journal of Neural Engineering*, 8(4), 046023. http://dx.doi.org/
10.1088/1741-2560/8/4/046023, S1741-2560(11)72859-7 [pii].

Lu, M. C., Tsai, C. C., Chen, S. C., Tsai, F. J., Yao, C. H., & Chen, Y. S. (2009). Use of
electrical stimulation at different current levels to promote recovery after peripheral
nerve injury in rats. *Journal of Trauma*, 67(5), 1066–1072. http://dx.doi.org/10.1097/
TA.0b013e318182351a, 00005373-200911000-00027 [pii].

Mendonca, A. C., Barbieri, C. H., & Mazzer, N. (2003). Directly applied low intensity direct
electric current enhances peripheral nerve regeneration in rats. *Journal of Neuroscience
Methods*, 129(2), 183–190.

Ray, W. Z., & Mackinnon, S. E. (2010). Management of nerve gaps: Autografts, allografts,
nerve transfers, and end-to-side neurorrhaphy. *Experimental Neurology*, 223(1), 77–85.
http://dx.doi.org/10.1016/j.expneurol.2009.03.031, S0014-4886(09)00122-8 [pii].

Roman, G. C., Strahlendorf, H. K., Coates, P. W., & Rowley, B. A. (1987). Stimulation of
sciatic nerve regeneration in the adult rat by low-intensity electric current. *Experimental
Neurology*, 98(2), 222–232.

Singh, B., Xu, Q. G., Franz, C. K., Zhang, R., Dalton, C., Gordon, T., et al. (2012). Accelerated axon outgrowth, guidance, and target reinnervation across nerve transection gaps
following a brief electrical stimulation paradigm. *Journal of Neurosurgery*, 116(3), 498–512.
http://dx.doi.org/10.3171/2011.10.JNS11612.

Valero-Cabre, A., & Navarro, X. (2002). Functional impact of axonal misdirection after
peripheral nerve injuries followed by graft or tube repair. *Journal of Neurotrauma*,
19(11), 1475–1485. http://dx.doi.org/10.1089/089771502320914705.

Vivo, M., Puigdemasa, A., Casals, L., Asensio, E., Udina, E., & Navarro, X. (2008). Immediate electrical stimulation enhances regeneration and reinnervation and modulates
spinal plastic changes after sciatic nerve injury and repair. *Experimental Neurology*,
211(1), 180–193. http://dx.doi.org/10.1016/j.expneurol.2008.01.020, S0014-4886
(08)00051-4 [pii].

Wan, L. D., Xia, R., & Ding, W. L. (2010). Electrical stimulation enhanced remyelination of
injured sciatic nerves by increasing neurotrophins. *Neuroscience*, 169(3), 1029–1038.
http://dx.doi.org/10.1016/j.neuroscience.2010.05.051, S0306-4522(10)00791-8 [pii].

Wang, W. J., Zhu, H., Li, F., Wan, L. D., Li, H. C., & Ding, W. L. (2009). Electrical stimulation promotes motor nerve regeneration selectivity regardless of end-organ connection. *Journal of Neurotrauma*, 26(4), 641–649.

Wenjin, W., Wenchao, L., Hao, Z., Feng, L., Yan, W., Wodong, S., et al. (2011). Electrical
stimulation promotes BDNF expression in spinal cord neurons through Ca(2+)- and
Erk-dependent signaling pathways. *Cellular and Molecular Neurobiology*, 31(3),
459–467. http://dx.doi.org/10.1007/s10571-010-9639-0.

Yao, C. H., Chang, R. L., Chang, S. L., Tsai, C. C., Tsai, F. J., & Chen, Y. S. (2012). Electrical stimulation improves peripheral nerve regeneration in streptozotocin-induced diabetic rats. *Journal of Trauma and Acute Care Surgery*, 72(1), 199–205. http://dx.doi.org/
10.1097/TA.0b013e31822d233c, 01586154-201201000-00028 [pii].

CHAPTER SIX

Role of Physical Exercise for Improving Posttraumatic Nerve Regeneration

Paulo A.S. Armada-da-Silva[*,†,1], **Cátia Pereira**[*,†], **Sandra Amado**[†,‡],
António P. Veloso[*,†]

[*]Faculdade de Motricidade Humana, Universidade de Lisboa, Cruz Quebrada-Dafundo, Portugal
[†]Centro Interdisciplinar para o Estudo da Performance Humana (CIPER), Faculdade de Motricidade Humana, Cruz Quebrada-Dafundo, Portugal
[‡]UIS—Unidade de Investigação em Saúde, Escola Superior de Saúde de Leiria, Instituto Politécnico de Leiria, Portugal
[1]Corresponding author: e-mail address: parmada@fmh.ulisboa.pt

Contents

Abstract

Despite the great regenerative ability of the peripheral nervous system (PNS), traumatic peripheral nerve damage often causes severe chronic disability. Rehabilitation following PNS trauma usually employs therapeutic exercise in an attempt to reanimate the target organs and stimulate functional recovery. Over the past years, important neurobiological determinants of PNS regeneration and successful end-organ reinnervation were unveiled. Such knowledge provides cues for designing novel strategies for treating and rehabilitating traumatic PNS damage. Physical exercise, by means of treadmill or wheel running, is neuroprotective and neuroregenerative. Research conducted on rodents demonstrates that endurance exercise modulates several of the cellular and molecular responses to peripheral nerve injury and by doing so it stimulates nerve regeneration and functional recovery following experimental PNS injury. Treadmill running increases the number of regenerating neurons, the rate of axonal growth, and the

International Review of Neurobiology, Volume 109
ISSN 0074-7742
http://dx.doi.org/10.1016/B978-0-12-420045-6.00006-7

125

extent of muscle reinnervation following peripheral nerve injury. Furthermore, treadmill running has the ability to increase the release of neurotrophins and growth factors in the spinal cord, the injured nerve, and reinnervating muscles. Treadmill running also seems to prevent the development of neuropathic pain and allodynia as a result of peripheral nerve damage. In addition, physical exercise, even if performed for a short period of time, exerts positive conditioning effects in neuroregeneration capacity, improving the acute response to peripheral nerve insults. Some of these effects can also be obtained with passive exercise or manual stimulation. In humans, however, evidence demonstrating a positive effect of exercise on nerve regeneration is at best poor.

1. INTRODUCTION

Despite advances in treatment, traumatic peripheral nerve injuries remain an important cause of chronic disability and pain with a negative impact in health-related quality of life (Ciaramitaro et al., 2010). The health and social burden of peripheral nervous system (PNS) trauma is made worse by its higher prevalence in young adults, as a result of traffic and sport-related accidents (Lad, Nathan, Schubert, & Boakye, 2010).

Several neurobiological mechanisms underlying PNS regenerations and reinnervation of target organs have been identified in the past years (de Ruiter et al., 2008; Gordon, Tyreman, & Raji, 2011; Ijkema-Paassen, Meek, & Gramsbergen, 2002). Reinnervation and functional recovery success following peripheral nerve injury are affected by the distance from the site of injury to the denervated targets, which determines the duration of muscle denervation (Kobayashi et al., 1997), sluggish and staggered axonal elongation past the injury site (Al-Majed, Neumann, Brushart, & Gordon, 2000; Brushart, 1990), loss of regeneration support by the distal nerve and by denervated Schwann cells (Gordon et al., 2011), inaccurate reinnervation of the targets (Angelov et al., 2007; de Ruiter et al., 2008), and severe muscle atrophy and muscle fiber degeneration (Dedkov, Kostrominova, Borisov, & Carlson, 2001; Gordon et al., 2011). Neuronal cell death is possibly another potential factor compromising functional outcome after traumatic PNS injury, particularly if the injury is near the central nervous system (CNS) and involves sensory neurons (Abrams & Widenfalk, 2005).

Physical exercise, by means of voluntary or forced locomotion, is one of several possible strategies to enhance peripheral nerve regeneration and improve target reinnervation (English, Wilhelm, & Sabatier, 2011; Udina, Cobianchi, Allodi, & Navarro, 2011). Although therapeutic exercise stands as a common practice in the rehabilitation of PNS lesions for many decades, only recently was it shown that physical exercise and activity-dependent interventions have real impact on neurobiological mechanisms of peripheral nerve regeneration. Here, we will review research conducted over the past years to investigate the role of walking or running exercise on nerve regeneration and reinnervation following peripheral nerve injury, mainly conducted in rodents. The role of the walking/running exercise will be analyzed within the context of the cellular and molecular events taking place during nerve healing. We will also address other kinds of movement stimulation, in particular, the role of manual mobilization of the denervated limbs and how this may affect nerve regeneration. We will end this review by providing a brief synthesis of the main conclusions gathered from clinical research that assessed the efficacy of physical training and other forms of activity-related treatment in treating and rehabilitating people with peripheral neuropathy.

2. THE EFFECT OF EXERCISE TRAINING ON NERVE REGENERATION

2.1. Early studies

Initial studies addressing the effect of physical exercise on nerve regeneration and functional recovery following PNS injury found conflicting results (reviewed in van Meeteren, Brakkee, Hamers, Helders, & Gispen, 1997). These studies were conducted in the rat and in the mouse, and occasionally in other animal models, and employed endurance exercise (Gutmann & Jakoubek, 1963; Herbison, Jaweed, & Ditunno, 1974; van Meeteren, Brakkee, Helders, & Gispen, 1998) or muscle overloading by ablation of synergistic muscles (Herbison, Jaweed, & Ditunno, 1973), to stimulate nerve regeneration and end-organ reinnervation, usually after sciatic nerve crush injury. The exercise protocols consisted on either spontaneous exercise, typically voluntary wheel running (Irintchev, Draguhn, & Wernig, 1990), or forced exercise, especially swimming (Gutmann & Jakoubek, 1963; Herbison et al., 1974; van Meeteren et al., 1998) and treadmill running (Herbison, Jaweed, & Ditunno, 1980).

Treadmill running or swimming failed to improve muscle reinnervation and function in several occasions (Herbison et al., 1974, 1980; Irintchev et al., 1990; van Meeteren et al., 1998), probably due to excessive physical exertion. Accordingly, increasing muscular activity by tenotomy of synergists or by swimming exercise was reported to cause considerable muscle damage (Herbison, Jaweed, Ditunno, & Scott, 1973), especially if applied too early after PNS injury (Herbison et al., 1974). The impact of physical exercise in the function of reinnervated muscles is not straightforward. Free access to running wheels by mice following tibial nerve transection and repair caused delayed reinnervation of soleus muscle but strengthened synaptic transmission at the level of the neuromuscular junction, at least during the initial weeks following the nerve injury (Badke, Irintchev, & Wernig, 1989). This is in accordance with findings demonstrating inhibition of nerve sprouting in muscle partially denervated by L4 or L5 ventral root avulsion due to running exercise (Tam, Archibald, Jassar, Tyreman, & Gordon, 2001). The conflicting results regarding the effect of physical exercise on nerve regeneration and functional recovery probably reflect differences in the type, duration, and intensity of the exercise regimens. In fact, treadmill running following sciatic nerve crush in the rat impaired recovery of motor function, but not of sensitive function (van Meeteren et al., 1998), whereas swimming or rising on the rear limbs for water reaching globally improved functional recovery (van Meeteren et al., 1997).

2.2. Physical exercise action on neurobiological mechanisms of nerve regeneration

Although physical exercise may aid recovery after PNS damage or act in the opposite way, the view that physical exercise stimulates key neurobiological processes underlying nerve regeneration is getting wide acceptance. Table 6.1 summarizes evidence regarding the effect of exercise stimulation on a number of molecular mechanisms of nerve regeneration. Retrograde labeling of motoneurons (English, Cucoranu, Mulligan, & Sabatier, 2009), direct visualization of fluorescent growing axons (Sabatier, Redmon, Schwartz, & English, 2008), and counting of nerve fibers in semithin transverse nerve sections (Asensio-Pinilla, Udina, Jaramillo, & Navarro, 2009; Ilha et al., 2008), combined with neurophysiological and behavioral outcome measures, support the view that physical exercise promotes nerve regeneration and functional recovery in models of traumatic nerve injury (Udina, Cobianchi, et al., 2011; reviewed in English, Wilhelm, et al., 2011).

2.2.1 Physical exercise, neuronal cell survival, and neuroprotection

Critical for successful nerve regeneration is ensuring survival of axotomized neurons, together with increasing the number of growing neurons. Several studies demonstrate that treadmill exercise augments the number of axons

Table 6.1 Major neurotrophic factors affected by physical exercise

Neurotrophic factor	Receptor	Role in neuroprotection and neuroregeneration	Pattern of expression	Expression during peripheral nerve regeneration	Effect of physical activity	References
BDNF	trkB (high affinity) p75^{NTR} (low affinity)	Neuronal survival and neuroprotection Axonal growth and neurofilament assembly Axonal elongation Promotion of myelination Synaptic plasticity	Differently expressed in motor and sensory neurons Production and secretion from several cell types (neurons, Schwann cells, and muscle fibers)	Expression of BDNF molecule and of its receptors is upregulated after peripheral nerve injury BDNF molecule and its receptors are expressed at the axonal ends and Schwann cells	BDNF levels (protein and mRNA) increase in brain and spinal cord with physical exercise and in proportion to the distance run BDNF muscle production is upregulated by muscle contraction (including active exercise) BDNF levels increase in both injured and intact nerves with physical exercise (intact nerve: protein and mRNA levels)	Ying, Roy, Edgerton, and Gomez-Pinilla (2005) Gomez-Pinilla, Ying, Opazo, Roy, and Edgerton (2001) Gomez-Pinilla et al. (2001), Vaynman, Ying, and Gomez-Pinilla (2003), Ying et al. (2005)

Continued

Table 6.1 Major neurotrophic factors affected by physical exercise—cont'd

Neurotrophic factor	Receptor	Role in neuroprotection and neuroregeneration	Pattern of expression	Expression during peripheral nerve regeneration	Effect of physical activity	References
					Expression of trkB receptors is increased by exercise BDNF and trkB expression is required for the nerve-regenerating effect of active and passive exercises	Ghiani et al. (2007), Qin et al. (2006), Wood et al. (2011) Sohnchen et al. (2010), Wilhelm et al. (2012)
NT-3	trkC (high affinity) p75^NTR (low affinity)	Neuronal survival Axonal sprouting Axonal elongation Possible contribution for preferential motor reinnervation Synaptogenesis and synaptic plasticity Maintenance of muscle trophism	Present in higher concentrations in sensory than in motor nerves NT-3 is predominantly expressed and produced by muscle trkC is abundantly expressed in motoneurons	NT-3 is produced by Schwann cells in injured nerves (likely main source of NT-3 during the initial nerve response to injury)	NT-3 and trkC are upregulated by physical exercise in both intact spinal cord and muscle	Gomez-Pinilla et al. (2001), Ying, Roy, Edgerton, and Gomez-Pinilla (2003)

NT-4/5	trkB (high affinity) p75^{NTR} (low affinity)	Survival of motoneurons Axonal sprouting in adult intact nerve NMJ functional plasticity Preferential recovery of slow motor units	Preferential production by slow-type muscle fibers	NT-4/5 muscle expression declines after nerve injury NT-4/5 levels increase with electrical stimulation	NT-4/5 muscle levels decline with muscle denervation	Funakoshi et al. (1995)
					NT-4/5 muscle levels increase with muscle activity	Funakoshi et al. (1995)
					NT-4/5 muscle levels do not change with mechanical compression, passive stretching, or muscle injury	Funakoshi et al. (1995)
					NT-4/5 expression in proximal nerve stumps is required for the effect of treadmill running in enhancing axonal growth	English, Meador, and Carrasco (2005)
					Expression of trkB receptors is upregulated by exercise	Ghiani, Ying, Vellis, and Gomez-Pinilla (2007), Wood et al. (2011)

Continued

Table 6.1 Major neurotrophic factors affected by physical exercise—cont'd

Neurotrophic factor	Receptor	Role in neuroprotection and neuroregeneration	Pattern of expression	Expression during peripheral nerve regeneration	Effect of physical activity	References
HGF	c-met	Survival of subpopulations of motoneurons			Possibly supports activity-dependent motoneuronal survival	Yamamoto et al. (1997)
IGF-1	IGFR-1	Survival and neuroprotection of motoneurons Branching of regenerating motoneurons		IGF-1 is overexpressed by denervated Schwann cells	Supports activity-related motoneuronal survival (intact nerve)	Deforges et al. (2009)
					IGF-1 required for passive exercise-induced functional recovery after facial nerve damage	Kiryakova et al. (2010)

GDNF	GFRα1 (high affinity) c-ret	Survival of motoneurons Axonal elongation Proliferation and myelination of Schwann cells Formation, maintenance, and plasticity of NMJ	Expressed in both sensory and motor neurons and muscle (mRNA)	GDNF and GFRα1 increase in the distal nerve stump (Schwann cells) and muscles after nerve injury	GDNF production and release by denervated and reinnervated muscles	McCullough, Peplinski, Kinnell, and Spitsbergen (2011)
				GFRα1 is expressed at motor nerve terminals GFRα1 is deleterious after long-term denervation	GDNF muscle levels increase with muscle activity (passive and active exercise) and decrease with immobilization	McCullough et al. (2011), Wehrwein, Roskelley, and Spitsbergen (2002)
				GFRα1 concentration at postsynaptic membrane regulates sprouting in reinnervated muscles	GDNF muscle levels decrease with augmented release of acetylcholine	McCullough et al. (2011)
					GDNF levels decrease in the dorsal horn with treadmill exercise	Cobianchi, Casals-Diaz, Jaramillo, & Navarro (2013)

BDNF, brain–derived neurotrophic factor; c-met, hepatocyte growth factor receptor; c-ret, Ret proto-oncogene; GDNF, glial-derived neurotrophic factor; GFRα1, GDNF family receptor α1; IGF-1, insulin–like growth factor 1; IGF-2, insulin–like growth factor 2; IGFR–1, insulin–like growth factor receptor 1; NMJ, neuromuscular junction; p75^{NTR}, p75 neurotrophin receptor; trkB, tyrosine kinase receptor B; trkC, tyrosine kinase receptor C.

that regenerate across the injury site, either in rats (Seo et al., 2006) or in mice (English, Schwartz, Meador, Sabatier, & Mulligan, 2007; Sabatier et al., 2008). The release of distinct neurotrophic factors within the spinal cord is critical for survival of motoneurons. Although a clear relationship between physical exercise and increased production and release of factors, such as of leukemia inhibitory factor and ciliary neurotrophic factor (Dahlin & Brandt, 2004), is lacking, other neuronal factors known to promote neuronal survival are upregulated by increased levels of physical activity. Some of these include neurotrophin (NT)-3 (Gomez-Pinilla, Ying, Opazo, Roy, & Edgerton, 2001; Sendtner, Pei, Beck, Schweizer, & Wiese, 2000; Sterne, Coulton, Brown, Green, & Terenghi, 1997; Ying et al., 2003) and NT-4/5, glial-derived neurotrophic factor (GDNF) (Grumbles, Sesodia, Wood, & Thomas, 2009; Sendtner et al., 2000), insulin-like growth factor 1 (IGF-1) and IGF-2, and the brain-derived neurotrophic factor (BDNF) (Gomez-Pinilla et al., 2001; Wilhelm et al., 2012; Yarrow, White, McCoy, & Borst, 2010; Ying et al., 2005).

Motoneuron death is relatively common early during development. In this case, GDNF is a potent rescue factor from developmentally programmed and axotomy-related neuronal death (reviewed in McCullough et al., 2011). GDNF is produced by Schwann cells and skeletal muscle, namely, after denervation (McCullough et al., 2011; Vianney & Spitsbergen, 2011). Levels of GDNF at the spinal cord tend to decrease as a consequence of peripheral nerve injury. While treadmill exercise alone cannot revert such decrease, when combined with electrical stimulation, it significantly raises GDNF levels in the spinal cord following sciatic nerve cut and repair (Cobianchi et al., 2013).

BDNF is particularly relevant in what concerns the effect of physical exercise on the brain, spinal cord, and PNS. BDNF is produced and secreted by several cell types including neurons, Schwann cells, and muscle fibers (Koppel et al., 2009). Physical exercise stimulates both nerve- and muscle-derived BDNF release and increases it circulating levels in humans (Gomez-Pinilla et al., 2001; Rojas Vega et al., 2006; Yarrow et al., 2010; Ying et al., 2003). However, the BDNF system is not always neuroprotective. Unlike mature BDNF, proBDNF protein, acting through the $p75^{NTR}$ neurotrophic receptor, is proapoptotic and causes synaptic regression (Teng et al., 2005). Importantly, wheel running enhances the synthesis of mature BDNF from proBDNF in mice, through activation of tissue-type plasminogen activator (Ding, Ying, & Gomez-Pinilla, 2011).

2.2.2 Axonal regeneration and rate of elongation

Two major contributions of physical exercise to peripheral nerve regeneration are the increase in the number of outgrowing axons and the increase in the rate of axonal elongation (English, Cucoranu, et al., 2011; English et al., 2009; English, Wilhelm, et al., 2011; English, Cucoranu, et al., 2011; English, Wilhelm, et al., 2011; Sabatier et al., 2008). The studies of Doyle and Roberts (2006), in the spinal cord, and of Sabatier et al. (2008), in the sciatic nerve, are two exemplar studies that demonstrate the ability of physical exercise in increasing the rate of elongation of regenerating axons. In the latter study, different treadmill running protocols were used, ranging from continuous 1-h exercise (belt speed: 10 m/min) to low- and high-intensity interval training, with different number of repetitions (belt speed: 20 m/min) (Sabatier et al., 2008). In general, treadmill exercise during the initial 2 weeks of recovery accelerated the growth of axons in the common fibular nerve of the *thy-1-YFP-H* mice. Continuous 1-h treadmill running and high-intensity interval training both were associated with axonal profiles that were twice the length of those observed in unexercised animals (Sabatier et al., 2008). In another study, mild-intensity continuous (1 h, 10 m/min) or interval (20 m/min, 2 min exercise, 5 min rest) treadmill exercise for 2 weeks immediately following sciatic nerve transection and repair increased the number of regenerating motoneurons, compared to unexercised animals (English et al., 2009). Importantly, the effect of treadmill running in stimulating regeneration of axotomized motoneurons was of similar magnitude of that seen with electrical stimulation or treatment with the glycosaminoglycans-degrading enzyme chondroitinase ABC, but avoiding the undesired axonal misrouting (English et al., 2009; Sabatier et al., 2008). Treadmill walking/running also raises the number of myelinated nerve fibers (Asensio-Pinilla et al., 2009; Ilha et al., 2008; Udina, Puigdemasa, & Navarro, 2011) and, in some cases, improves myelin sheath thickness of regenerated peripheral nerves (Ilha et al., 2008).

The effect of treadmill running in increasing the elongation rate of regenerating axons relies on expression of BDNF (Molteni, Zheng, Ying, Gómez-Pinilla, & Twiss, 2004; Wilhelm et al., 2012) and NT-4/5 (English, Cucoranu, et al., 2011; English, Wilhelm, et al., 2011) in axons and Schwann cells. Mice of the *thy-YFP-H* lineage, lacking BDNF in their YFP+ neurons, show defective ability to regenerate their axons though nerve grafts harvested from wild-type counterparts or from mice also lacking the expression of BDNF by Schwann cells (Wilhelm et al., 2012). Treadmill running was unable to correct the abnormal axonal growth of neurons not expressing BDNF (Wilhelm et al., 2012). However, in animals normally expressing BDNF, treadmill exercise was capable of enhancing axonal elongation even through nerve grafts void of this neurotrophin (Wilhelm et al., 2012). Likewise, NT-4/5 expression within the regenerating nerve is required for treadmill running to enhance axonal growth (English, Cucoranu, et al., 2011; English, Wilhelm, et al., 2011). In NT-4/5 knockout mice, treadmill running is unable to alter the rate at which axons grow along the common fibular nerve. However, nerve grafts from NT-4/5 knockout donors or acellular nerve grafts do not cancel the role played by treadmill running in stimulating axonal growth, thereby indicating that the NT-4/5 acts in the proximal nerve stump, most likely at the level of the regenerating neurons (English, Cucoranu, et al., 2011; English, Wilhelm, et al., 2011).

2.2.3 Improving the nerve regenerative environment

The distal nerve pathway plays a key role in sustaining peripheral nerve regeneration. Schwann cells provide guidance to regenerating axons (Allodi, Udina, & Navarro, 2012), and prolonged denervation of the distal nerve stump is a major reason for impaired reinnervation of target organs (Gordon et al., 2011; Höke & Brushart, 2010). Only few studies exist relating physical exercise and Schwann cell function in axotomized peripheral nerves. Seo et al. (2006) reported that treadmill exercise, undertaken during the initial 14 days after sciatic nerve injury, improves sciatic nerve regeneration and functional recovery. In this case, treadmill exercise increased the levels of cell division cycle 2 (cdc2) mRNA and protein at the injury site. This marker of tissue regeneration colocalized with the Schwann cell marker S100 beta protein, and treatment with a cdc2 antagonist counteracted the effect of treadmill exercise in improving nerve regeneration (Seo et al., 2006). In a sequence study, mild-intensity treadmill exercise also stimulated Schwann cell proliferation following sciatic nerve injury in the rat via increase of extracellular regulated kinase (Erk) 1/2 activity. In this study, phosphorylated Erk1/2 upregulation in Schwann cells followed the increase of growth-associated protein (GAP)-43 mRNA and protein levels in the sciatic nerve, as well as the increase of the phosphorylated form of the downstream effector c-Jun (Seo et al., 2009). Another sort of mechanisms that may promote Schwann cell function is related with preventing cell apoptosis in the distal nerve. In a study undertaken by Shokouhi et al. (2008), treadmill running exerted a protective effect against lipid peroxidation in intact nerves, thus reducing Schwann cell apoptosis. A further potential mechanism by which physical exercise might help peripheral nerve regeneration is by diminishing the amount of myelin-related inhibitors of axonal growth (Filbin, 2003; Gaudet, Popovich, & Ramer, 2011). In mice, wheel running decreased the levels of myelin-associated glycoprotein in the spinal cord, while enhanced neurite outgrowth of cortical neurons growing on myelin is collected from the running-trained animals (Ghiani et al., 2007). The lowering action of running exercise on the levels of myelin-associated glycoprotein in the spinal cord seems to be regulated by BDNF and protein kinase A (Ghiani et al., 2007).

2.2.4 Enhanced nerve regeneration by conditioning exercise

Physical exercise seems to exert a conditioning effect over the nervous system that potentiates its ability to respond to subsequent damage (Bobinski et al., 2011; Molteni et al., 2004). Voluntary wheel running for a period of just 3 and 7 days clearly enhanced the rate of neurite outgrowth in cultured exercise-conditioned dorsal root ganglion neurons as well as increased the number of sensory neurons regenerating following sciatic nerve injury (Molteni et al., 2004). The conditioning effect of wheel running was associated with higher BDNF and NT-3 expression in sensory ganglia of sciatic-injured animals as well as with increased levels of synapsin I and GAP-43. Such conditioning effect of treadmill exercise relied on neurotrophins signaling through tyrosine kinase receptors (Molteni et al., 2004).

Treadmill exercise conducted prior to sciatic nerve crush injury was also reported to improve recovery of motor and sensitive function in mice, while treadmill exercise performed both before and after the sciatic nerve injury resulted in better morphology of the regenerated sciatic nerves, with increased number of myelinated fibers and increased myelin thickness (Bobinski et al., 2011). Together with the positive role on functional recovery, conditioning treadmill exercise blunted the response of interleukin (IL)-1β in both the regenerating sciatic nerve and the spinal cord, and of IL-6 receptor in the spinal cord, but not in the sciatic nerve (Bobinski et al., 2011).

2.3. Effect of passive physical exercise on nerve regeneration and functional recovery

Passive exercise can be employed in the early stages of rehabilitation of PNS injuries, commonly through manual limb mobilization. In an early study, passive mobilization of the hindpaw, conducted before muscle reinnervation, was able to stimulate nerve sprouting, although it had no effect

to structural changes in nerves and diminish their ability to sustain physiologic strain. Accordingly, motor end plate degeneration and polyinnervation (Pachter & Eberstein, 1984), delayed soleus reinnervation (Marciniak, 1985), and deepest loss of contractile force during muscle reinnervation (Herbison et al., 1984) accompany limb immobilization.

There is growing acceptance that mechanical strain plays a role in growth, guidance, and function of neurons (reviewed in Suter & Miller, 2011). Axons display a remarkable ability to elongate in response to externally applied forces. Cultured neurons can grow their axons at an extraordinary rate of 8 mm/day (Pfister, Iwata, Meaney, & Smith, 2004). Of course, such high axonal elongation rate cannot be directly translated to *in vivo* conditions. However, it has been shown that extreme leg-lengthening causes minute axon damage in the rat's sciatic nerve (Abe et al., 2004). Although current understanding of the mechanisms of rapid growth of stretched axons is still incomplete, it is nevertheless certain that it relies on intracellular signaling set in motion by mechanotransduction (Pfister et al., 2004). That passive exercise may potentially trigger these signaling pathways is an attractive hypothesis.

3. EFFECT OF EXERCISE ON NEUROPATHIC PAIN

Neuropathic pain is a frequent and adverse consequence of traumatic PNS injury and is a major cause of disability and poor quality of life in these conditions (Ciaramitaro et al., 2010). Treadmill exercise has been shown to reduce allodynia and hyperalgesia due to chronic constriction injury (CCI) (Cobianchi, Marinelli, Florenzano, Pavone, & Luvisetto, 2010) or sciatic nerve crush (Bobinski et al., 2011). The impact of treadmill exercise on the development of allodynia is dependent on the timing and duration of the exercise protocol. Accordingly, treadmill running for 1 h/day from day 3 to day 7 following CCI diminishes signs of allodynia throughout the entire survival time and improves functional recovery, while keeping the exercise protocol for a longer period of time produces the opposite effect (Cobianchi et al., 2010). The antinociceptive effect of short-term treadmill running is in relationship with blunted microglia and astrocytes activation in the dorsal and ventral horns of the spinal cord ipsilateral to the constricted sciatic nerve (Cobianchi et al., 2010). Mechanisms relating treadmill running and diminished neuropathic pain probably involve changed cytokines and that of proinflammatory mediators release, within the spinal cord (Bobinski et al., 2011), and modulation of the spatial and temporal pattern of neurotrophins expression both in the CNS and in the PNS (Cobianchi et al., 2013). Accordingly, in female rats, 1 h daily of treadmill running at mild to moderate intensity, performed from day 3 to day 7 after sciatic nerve transection and direct repair, reduces nociception in the hindpaw and improves response latencies in the affected hindlimb to either thermal or mechanical stimulation (Cobianchi et al., 2013). Lower expression of pronociceptive BDNF and GDNF in the dorsal horn accompanies treadmill running-associated hypoalgesia (Cobianchi et al., 2013).

Joint manual mobilization also reduces the intensity of ankle joint pain and the extent of activity along nociceptive pathways in the spinal cord that are induced experimentally by capsaicin injection into the ankle joint (Malisza et al., 2003). In addition, ankle mobilization applied every other day for the first 5 weeks following sciatic nerve crush in the rat reduces hyperalgesia and spinal glial activation and accelerates normalization of sciatic function index scores (Martins et al., 2011). The hypoalgesic property of physical exercise and joint movement is not specific to neuropathic pain and appears to involve the opioid and serotonergic systems (Mazzardo-Martins et al., 2010).

4. TRANSLATIONAL RESEARCH AND CLINICAL STUDIES

In relative contrast with research on animal models, clinical research fails in demonstrating the therapeutic potential of physical exercise and passive mobilization in the treatment of traumatic PNS injury in human subjects. From Cochrane Database of systematic reviews, there is little evidence that exercise therapy contributes to ameliorate functional ability in people with neuropathy (White, Pritchard, & Turner-Stokes, 2004), carpal tunnel syndrome (Page, O'Connor, Pitt, & Massy-Westropp, 2012), ulnar neuropathy (Caliandro, La Torre, Padua, Giannini, & Padua, 2012), or Bell's palsy (Teixeira, Valbuza, & Prado, 2011). In addition, endurance exercise, meaning low- to moderate-intensity prolonged exercise, seems to bring no additional benefit in measures of functional ability in participants with peripheral neuropathy (White et al., 2004). However, there is some evidence that resistance exercise programs, thus more focused in stimulating force generation by specific muscle groups, moderately improve muscle strength in people with PNS dysfunction (White et al., 2004).

Few studies have been conducted in facial palsy and carpal tunnel syndrome patients to assess the efficacy of exercise interventions. From a single small study, there is evidence that active facial exercises decrease disability in people with chronic facial palsy and diminished the prevalence of synkinesis in these patients (Teixeira et al., 2011). As for the carpal tunnel syndrome, a systematic review of the impact of different kinds of passive mobilization on a number of outcomes, ranging from functional ability, quality of life, and imaging and neurophysiological assessments, negates the effectiveness of such interventions in resolving this condition (Page et al., 2012). The discrepancy between experimental animal studies and clinical research regarding the effect of physical exercise on PNS regeneration is likely related to the complexity of clinical cases of traumatic PNS injury, together with the difficulty in standardizing study participants and exercise protocols in clinical and translational research.

5. CONCLUSIONS

In conclusion, research conducted over the past decades has provided convincing evidence of the neuroregenerative role of physical exercise. Figure 6.1 illustrates some of the demonstrated actions of treadmill running at different places, including the CNS and the PNS and reinnervating muscles, and several processes underlying neuroregeneration and neuroplasticity. Increased locomotor activity is conceived as a stimulus that increases neural activity in the spinal cord and axotomized neurons through increased drive from descending pathways and from peripheral inputs, either from regenerating sensory neurons or through propriospinal neural circuitry. This more physiological stimulation enhances axotomized neurons' sprouting and growth and also aids in preventing secondary and clinically relevant conditions, like neuropathic pain. Factors such as exercise intensity, duration, and timing for initiating the exercise treatment may decisively alter the effectiveness of physical exercise in improving outcomes after traumatic PNS injury. In rodent models of PNS injury, manual stimulation is effective in improving regeneration and functional recovery. In this case, manual stimulation, or other kinds of passive exercise, seems to share the effects of active, endurance exercise, with the advantage of being a treatment modality that can be employed early following the traumatic PNS injury, while also inducing less mechanical and physiological stresses. Only few clinical studies investigated the role of endurance exercise, resistance training, or

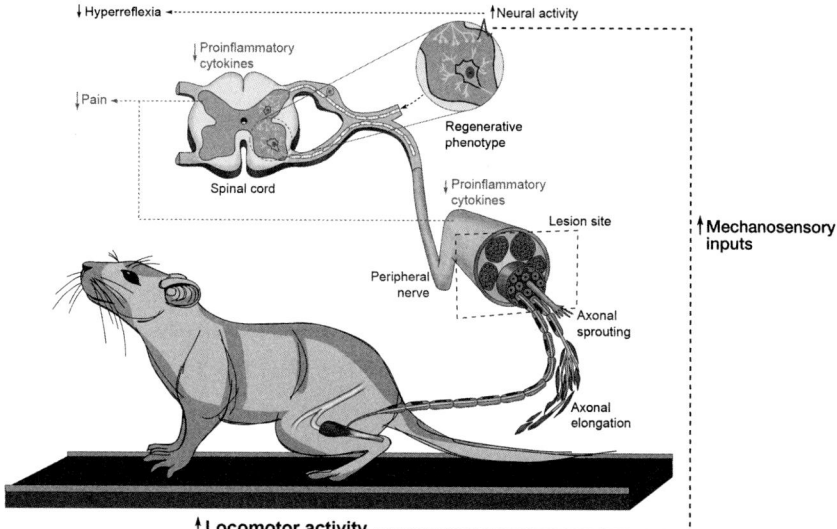

Figure 6.1 Schematic illustration of hypothetical mechanisms supporting the effect of physical exercise on nerve regeneration and functional recovery. Increased locomotion by means of treadmill exercise gives rise to mechanosensory inputs that converge to CNS and increase the input onto axotomized motoneurons, through regenerating sensory neurons or propriospinal circuitry. Activation of descendent spinal pathways also increases during treadmill running (not shown). Converging excitatory inputs enhance the regenerative response of axotomized motoneurons and promote the increase in the number of outgrowing axons into the distal nerve stump. Increased activity within neural pathways also stimulates axonal branching and raises the rate of axonal elongation. Treadmill running might also increase the levels of neurotrophic factors in the ventral horn of the spinal cord and in the regenerating nerve. The levels of proinflammatory cytokines in the spinal cord and injured peripheral nerve are lowered by treadmill exercise undertaken before and during the immediate days following PNS injury, reflecting either lower neuroinflammatory response magnitude or faster resolution of this response. Diminished levels of proinflammatory cytokines and of pronociceptive neurotrophic factors in the dorsal horn of the spinal cord (e.g., GDNF and NT-3), as a result of treadmill running, blunts glial activation, hyperreflexia, and prevents the development of neuropathic pain. Mechanosensory inputs delivered by passive exercise trigger a similar neuroprotective and neuroregenerative response. (For color version of this figure, the reader is referred to the online version of this chapter.)

manual mobilization in treating peripheral neuropathy in patients with results only moderately positive.

ACKNOWLEDGMENT

This work was partially supported by grant PTDC/DES/104036/2008 from Fundação para a Ciência e a Tecnologia, Ministério da Educação e Ciência, Portugal.

REFERENCES

Abe, I., Ochiai, N., Ichimura, H., Tsujino, A., Sun, J., & Hara, Y. (2004). Internodes can nearly double in length with gradual elongation of the adult rat sciatic nerve. *Journal of Orthopaedic Research, 22*(3), 571–577.

Abrams, M., & Widenfalk, J. (2005). Emerging strategies to promote improved functional outcome after peripheral nerve injury. *Restorative Neurology and Neuroscience, 23*(5–6), 367–382.

Allodi, I., Udina, E., & Navarro, X. (2012). Specificity of peripheral nerve regeneration: Interactions at the axon level. *Progress in Neurobiology, 98*(1), 16–37.

Al-Majed, A. A., Neumann, C. M., Brushart, T. M., & Gordon, T. (2000). Brief electrical stimulation promotes the speed and accuracy of motor axonal regeneration. *Journal of Neuroscience, 20*(7), 2602–2608.

Angelov, D. N., Ceynowa, M., Guntinas-Lichius, O., Streppel, M., Grosheva, M., Kiryakova, S. I., et al. (2007). Mechanical stimulation of paralyzed vibrissal muscles following facial nerve injury in adult rat promotes full recovery of whisking. *Neurobiology of Disease, 26*(1), 229–242.

Asensio-Pinilla, E., Udina, E., Jaramillo, J., & Navarro, X. (2009). Electrical stimulation combined with exercise increase axonal regeneration after peripheral nerve injury. *Experimental Neurology, 219*(1), 258–265.

Badke, A., Irintchev, A. P., & Wernig, A. (1989). Maturation of transmission in reinnervated mouse soleus muscle. *Muscle & Nerve, 12*(7), 580–586.

Benavides Damm, T., Richard, S., Tanner, S., Wyss, F., Egli, M., & Franco-Obregón, A. (2013). Calcium-dependent deceleration of the cell cycle in muscle cells by simulated microgravity. *The FASEB Journal, 27*(5), 2045–2054.

Bobinski, F., Martins, D. F., Bratti, T., Mazzardo-Martins, L., Winkelmann-Duarte, E. C., Guglielmo, L. G., et al. (2011). Neuroprotective and neuroregenerative effects of low-intensity aerobic exercise on sciatic nerve crush injury in mice. *Neuroscience, 194*, 337–348.

Brushart, T. M. (1990). Preferential motor reinnervation: A sequential double-labeling study. *Restorative Neurology and Neuroscience, 1*(3), 281–287.

Caliandro, P., La Torre, G., Padua, R., Giannini, F., & Padua, L. (2012). Treatment for ulnar neuropathy at the elbow. *Cochrane Database of Systematic Reviews, 7*, CD006839.

Caroni, P., Schneider, C., Kiefer, M. C., & Zapf, J. (1994). Role of muscle insulin-like growth factors in nerve sprouting: Suppression of terminal sprouting in paralyzed muscle by IGF-binding protein 4. *The Journal of Cell Biology, 125*(4), 893–902.

Cheema, U., Brown, R., Mudera, V., Yang, S. Y., McGrouther, G., & Goldspink, G. (2005). Mechanical signals and IGF-I gene splicing in vitro in relation to development of skeletal muscle. *Journal of Cellular Physiology, 202*(1), 67–75.

Cheng, H.-L., Randolph, A., Yee, D., Delafontaine, P., Tennekoon, G., & Feldman, E. L. (1996). Characterization of insulin-like growth factor-I and its receptor and binding proteins in transected nerves and cultured Schwann cells. *Journal of Neurochemistry, 66*(2), 525–536.

Ciaramitaro, P., Mondelli, M., Logullo, F., Grimaldi, S., Battiston, B., Sard, A., et al. (2010). Traumatic peripheral nerve injuries: Epidemiological findings, neuropathic pain and quality of life in 158 patients. *Journal of the Peripheral Nervous System, 15*(2), 120–127.

Clark, W. L., Trumble, T. E., Swiontkowski, M. F., & Tencer, A. F. (1992). Nerve tension and blood flow in a rat model of immediate and delayed repairs. *The Journal of Hand Surgery, 17*(4), 677–687.

Cobianchi, S., Casals-Diaz, L., Jaramillo, J., & Navarro, X. (2013). Differential effects of activity dependent treatments on axonal regeneration and neuropathic pain after peripheral nerve injury. *Experimental Neurology, 240*, 157–167.

Cobianchi, S., Marinelli, S., Florenzano, F., Pavone, F., & Luvisetto, S. (2010). Short- but not long-lasting treadmill running reduces allodynia and improves functional recovery after peripheral nerve injury. *Neuroscience, 168*(1), 273–287.

Coppieters, M. W., & Alshami, A. M. (2007). Longitudinal excursion and strain in the median nerve during novel nerve gliding exercises for carpal tunnel syndrome. *Journal of Orthopaedic Research, 25*(7), 972–980.

Dahlin, L. B., & Brandt, J. (2004). Basic science of peripheral nerve repair: Wallerian degeneration/growth cones. *Operative Techniques in Orthopaedics, 14*(3), 138–145.

Dedkov, E. I., Kostrominova, T. Y., Borisov, A. B., & Carlson, B. M. (2001). Reparative myogenesis in long-term denervated skeletal muscles of adult rats results in a reduction of the satellite cell population. *The Anatomical Record, 263*(2), 139–154.

Deforges, S., Branchu, J., Biondi, O., Grondard, C., Pariset, C., Lecolle, S., et al. (2009). Motoneuron survival is promoted by specific exercise in a mouse model of amyotrophic lateral sclerosis. *The Journal of Physiology, 587*(Pt. 14), 3561–3572.

de Ruiter, G. C., Malessy, M. J., Alaid, A. O., Spinner, R. J., Engelstad, J. K., Sorenson, E. J., et al. (2008). Misdirection of regenerating motor axons after nerve injury and repair in the rat sciatic nerve model. *Experimental Neurology, 211*(2), 339–350.

Dilley, A., Lynn, B., Greening, J., & DeLeon, N. (2003). Quantitative in vivo studies of median nerve sliding in response to wrist, elbow, shoulder and neck movements. *Clinical Biomechanics (Bristol, Avon), 18*(10), 899–907.

Ding, Q., Ying, Z., & Gomez-Pinilla, F. (2011). Exercise influences hippocampal plasticity by modulating brain-derived neurotrophic factor processing. *Neuroscience, 192*, 773–780.

Doyle, L. M. F., & Roberts, B. L. (2006). Exercise enhances axonal growth and functional recovery in the regenerating spinal cord. *Neuroscience, 141*(1), 321–327.

English, A. W., Cucoranu, D., Mulligan, A., Rodriguez, J. A., & Sabatier, M. J. (2011). Neurotrophin-4/5 is implicated in the enhancement of axon regeneration produced by treadmill training following peripheral nerve injury. *European Journal of Neuroscience, 33*(12), 2265–2271.

English, A. W., Cucoranu, D., Mulligan, A., & Sabatier, M. (2009). Treadmill training enhances axon regeneration in injured mouse peripheral nerves without increased loss of topographic specificity. *The Journal of Comparative Neurology, 517*(2), 245–255.

English, A. W., Meador, W., & Carrasco, D. I. (2005). Neurotrophin-4/5 is required for the early growth of regenerating axons in peripheral nerves. *European Journal of Neuroscience, 21*(10), 2624–2634.

English, A. W., Schwartz, G., Meador, W., Sabatier, M. J., & Mulligan, A. (2007). Electrical stimulation promotes peripheral axon regeneration by enhanced neuronal neurotrophin signaling. *Developmental Neurobiology, 67*(2), 158–172.

English, A. W., Wilhelm, J. C., & Sabatier, M. J. (2011). Enhancing recovery from peripheral nerve injury using treadmill training. *Annals of Anatomy, 193*(4), 354–361.

Evgenieva, E., Schweigert, P., Guntinas-Lichius, O., Pavlov, S., Grosheva, M., Angelova, S., et al. (2008). Manual stimulation of the suprahyoid-sublingual region diminishes polyinnervation of the motor endplates and improves recovery of function after hypoglossal nerve injury in rats. *Neurorehabilitation and Neural Repair, 22*(6), 754–768.

Filbin, M. T. (2003). Myelin-associated inhibitors of axonal regeneration in the adult mammalian CNS. *Nature Reviews. Neuroscience, 4*(9), 703–713.

Funakoshi, H., Belluardo, N., Arenas, E., Yamamoto, Y., Casabona, A., Persson, H., et al. (1995). Muscle-derived neurotrophin-4 as an activity-dependent trophic signal for adult motor neurons. *Science, 268*(5216), 1495–1499.

Gaudet, A. D., Popovich, P. G., & Ramer, M. S. (2011). Wallerian degeneration: Gaining perspective on inflammatory events after peripheral nerve injury. *Journal of Neuroinflammation, 8*, 110.

Ghiani, C. A., Ying, Z., Vellis, J. D., & GomezPinilla, F. (2007). Exercise decreases myelin-associated glycoprotein expression in the spinal cord and positively modulates neuronal growth. *Glia, 55*(9), 966–975.

Glazner, G. W., Morrison, A. E., & Ishii, D. N. (1994). Elevated insulin-like growth factor (IGF) gene expression in sciatic nerves during IGF-supported nerve regeneration. *Brain Research. Molecular Brain Research, 25*(3–4), 265–272.

Goldspink, G. (1999). Changes in muscle mass and phenotype and the expression of autocrine and systemic growth factors by muscle in response to stretch and overload. *Journal of Anatomy, 194*(3), 323–334.

Gomez-Pinilla, F., Ying, Z., Opazo, P., Roy, R. R., & Edgerton, V. R. (2001). Differential regulation by exercise of BDNF and NT-3 in rat spinal cord and skeletal muscle. *European Journal of Neuroscience, 13*(6), 1078–1084.

Gordon, T. (2009). The role of neurotrophic factors in nerve regeneration. *Neurosurgical Focus, 26*(2), E3.

Gordon, T., Tyreman, N., & Raji, M. A. (2011). The basis for diminished functional recovery after delayed peripheral nerve repair. *Journal of Neuroscience, 31*(14), 5325–5334.

Grothe, C., & Unsicker, K. (1987). Neuron-Enriched cultures of adult rat dorsal root ganglia: Establishment, characterization, survival, and neuropeptide expression in response to trophic factors. *Journal of Neuroscience Research, 18*(4), 539–550.

Grumbles, R. M., Sesodia, S., Wood, P. M., & Thomas, C. K. (2009). Neurotrophic factors improve motoneuron survival and function of muscle reinnervated by embryonic neurons. *Journal of Neuropathology and Experimental Neurology, 68*(7), 736–746.

Guntinas-Lichius, O., Hundeshagen, G., Paling, T., Streppel, M., Grosheva, M., Irintchev, A., et al. (2007). Manual stimulation of facial muscles improves functional recovery after hypoglossal-facial anastomosis and interpositional nerve grafting of the facial nerve in adult rats. *Neurobiology of Disease, 28*(1), 101–112.

Gutmann, E., & Jakoubek, B. (1963). Effect of increased motor activity on regeneration of the peripheral nerve in young rats. *Physiologia Bohemoslovaca, 12*, 463–468.

Herbison, G. J., Jaweed, M. M., & Ditunno, J. F., Jr. (1973). Reinnervating muscle in rats: The effect of overwork. *Archives of Physical Medicine and Rehabilitation, 54*(11), 511–514.

Herbison, G. J., Jaweed, M. M., & Ditunno, J. F. (1974). Effect of swimming on reinnervation of rat skeletal muscle. *Journal of Neurology, Neurosurgery & Psychiatry, 37*(11), 1247–1251.

Herbison, G. J., Jaweed, M. M., & Ditunno, J. F. (1980). Effect of activity and inactivity on reinnervating rat skeletal muscle contractility. *Experimental Neurology, 70*(3), 498–506.

Herbison, G. J., Jaweed, M. M., & Ditunno, J. F., Jr. (1984). Recovery of reinnervating rat muscle after cast immobilization. *Experimental Neurology, 85*(2), 239–248.

Herbison, G. J., Jaweed, M. M., Ditunno, J. F., & Scott, C. M. (1973). Effect of overwork during reinnervation of rat muscle. *Experimental Neurology, 41*(1), 1–14.

Höke, A., & Brushart, T. (2010). Introduction to special issue: Challenges and opportunities for regeneration in the peripheral nervous system. *Experimental Neurology, 223*(1), 1–4.

Ijkema-Paassen, J., Meek, M. F., & Gramsbergen, A. (2002). Reinnervation of muscles after transection of the sciatic nerve in adult rats. *Muscle & Nerve, 25*(6), 891–897.

Ilha, J., Araujo, R. T., Malysz, T., Hermel, E. E., Rigon, P., Xavier, L. L., et al. (2008). Endurance and resistance exercise training programs elicit specific effects on sciatic nerve regeneration after experimental traumatic lesion in rats. *Neurorehabilitation and Neural Repair, 22*(4), 355–366.

Irintchev, A., Draguhn, A., & Wernig, A. (1990). Reinnervation and recovery of mouse soleus muscle after long-term denervation. *Neuroscience, 39*(1), 231–243.

Kanje, M., Skottner, A., Sjöberg, J., & Lundborg, G. (1989). Insulin-like growth factor I (IGF-I) stimulates regeneration of the rat sciatic nerve. *Brain Research, 486*(2), 396–398.

Kim, B., Leventhal, P. S., Saltiel, A. R., & Feldman, E. L. (1997). Insulin-like growth factor-I-mediated neurite outgrowth in vitro requires mitogen-activated protein kinase activation. *Journal of Biological Chemistry, 272*(34), 21268–21273.

Kiryakova, S., Sohnchen, J., Grosheva, M., Schuetz, U., Marinova, T., Dzhupanova, R., et al. (2010). Recovery of whisking function promoted by manual stimulation of the vibrissal muscles after facial nerve injury requires insulin-like growth factor 1 (IGF-1). *Experimental Neurology, 222*(2), 226–234.

Kobayashi, J., Mackinnon, S. E., Watanabe, O., Ball, D. J., Gu, X. M., Hunter, D. A., et al. (1997). The effect of duration of muscle denervation on functional recovery in the rat model. *Muscle & Nerve, 20*(7), 858–866.

Koppel, I., Aid-Pavlidis, T., Jaanson, K., Sepp, M., Pruunsild, P., Palm, K., et al. (2009). Tissue-specific and neural activity-regulated expression of human BDNF gene in BAC transgenic mice. *BMC Neuroscience, 10*(1), 68.

Lad, S. P., Nathan, J. K., Schubert, R. D., & Boakye, M. (2010). Trends in median, ulnar, radial, and brachioplexus nerve injuries in the United States. *Neurosurgery, 66*(5), 953–960.

Malisza, K. L., Gregorash, L., Turner, A., Foniok, T., Stroman, P. W., Allman, A. A., et al. (2003). Functional MRI involving painful stimulation of the ankle and the effect of physiotherapy joint mobilization. *Magnetic Resonance Imaging, 21*(5), 489–496.

Marciniak, M. (1985). Regeneration of axonal endings of neuromuscular junctions of the soleus muscle in rats under conditions of hypodynamia. *Experimental Pathology, 27*(2), 111–118.

Martins, D. F., Mazzardo-Martins, L., Gadotti, V. M., Nascimento, F. P., Lima, D. A., Speckhann, B., et al. (2011). Ankle joint mobilization reduces axonotmesis-induced neuropathic pain and glial activation in the spinal cord and enhances nerve regeneration in rats. *Pain, 152*(11), 2653–2661.

Mazzardo-Martins, L., Martins, D. F., Marcon, R., dos Santos, U. D., Speckhann, B., Gadotti, V. M., et al. (2010). High-intensity extended swimming exercise reduces pain-related behavior in mice: Involvement of endogenous opioids and the serotonergic system. *The Journal of Pain: Official Journal of the American Pain Society, 11*(12), 1384–1393.

McCullough, M. J., Peplinski, N. G., Kinnell, K. R., & Spitsbergen, J. M. (2011). Glial cell line-derived neurotrophic factor protein content in rat skeletal muscle is altered by increased physical activity in vivo and in vitro. *Neuroscience, 174*, 234–244.

Molteni, R., Zheng, J.-Q., Ying, Z., Gómez-Pinilla, F., & Twiss, J. L. (2004). Voluntary exercise increases axonal regeneration from sensory neurons. *Proceedings of the National Academy of Sciences of the United States of America, 101*(22), 8473–8478.

Morcuende, S., Matarredona, E. R., Benítez-Temiño, B., Muñoz-Hernández, R., Pastor, Á. M., & de la Cruz, R. R. (2011). Differential regulation of the expression of neurotrophin receptors in rat extraocular motoneurons after lesion. *The Journal of Comparative Neurology, 519*(12), 2335–2352.

Navarro, X., Vivó, M., & Valero-Cabré, A. (2007). Neural plasticity after peripheral nerve injury and regeneration. *Progress in Neurobiology, 82*(4), 163–201.

Pachter, B. R., & Eberstein, A. (1984). Neuromuscular plasticity following limb immobilization. *Journal of Neurocytology, 13*(6), 1013–1025.

Pachter, B. R., & Eberstein, A. (1989). Passive exercise and reinnervation of the rat denervated extensor digitorum longus muscle after nerve crush. *American Journal of Physical Medicine & Rehabilitation, 68*(4), 179–182.

Page, M. J., O'Connor, D., Pitt, V., & Massy-Westropp, N. (2012). Exercise and mobilisation interventions for carpal tunnel syndrome. *Cochrane Database of Systematic Reviews, 6*, CD009899.

Pavlov, S. P., Grosheva, M., Streppel, M., Guntinas-Lichius, O., Irintchev, A., Skouras, E., et al. (2008). Manually-stimulated recovery of motor function after facial nerve injury requires intact sensory input. *Experimental Neurology, 211*(1), 292–300.

Pfister, B. J., Iwata, A., Meaney, D. F., & Smith, D. H. (2004). Extreme stretch growth of integrated axons. *The Journal of Neuroscience, 24*(36), 7978–7983.

Qin, D. X., Zou, X. L., Luo, W., Zhang, W., Zhang, H. T., Li, X. L., et al. (2006). Expression of some neurotrophins in the spinal motoneurons after cord hemisection in adult rats. *Neuroscience Letters*, *410*(3), 222–227.

Rojas Vega, S., Strüder, H. K., Vera Wahrmann, B., Schmidt, A., Bloch, W., & Hollmann, W. (2006). Acute BDNF and cortisol response to low intensity exercise and following ramp incremental exercise to exhaustion in humans. *Brain Research*, *1121*(1), 59–65.

Rydevik, B. L., Kwan, M. K., Myers, R. R., Brown, R. A., Triggs, K. J., Woo, S. L. Y., et al. (1990). An in vitro mechanical and histological study of acute stretching on rabbit tibial nerve. *Journal of Orthopaedic Research*, *8*(5), 694–701.

Sabatier, M. J., Redmon, N., Schwartz, G., & English, A. W. (2008). Treadmill training promotes axon regeneration in injured peripheral nerves. *Experimental Neurology*, *211*(2), 489–493.

Sendtner, M., Pei, G., Beck, M., Schweizer, U., & Wiese, S. (2000). Developmental motoneuron cell death and neurotrophic factors. *Cell and Tissue Research*, *301*(1), 71–84.

Seo, T. B., Han, I. S., Yoon, J. H., Hong, K. E., Yoon, S. J., & Namgung, U. (2006). Involvement of Cdc2 in axonal regeneration enhanced by exercise training in rats. *Medicine & Science in Sports & Exercise*, *38*(7), 1267–1276.

Seo, T. B., Oh, M. J., You, B. G., Kwon, K. B., Chang, I. A., Yoon, J. H., et al. (2009). ERK1/2-mediated Schwann cell proliferation in the regenerating sciatic nerve by treadmill training. *Journal of Neurotrauma*, *26*(10), 1733–1744.

Shokouhi, G., Tubbs, R. S., Shoja, M. M., Roshangar, L., Mesgari, M., Ghorbanihaghjo, A., et al. (2008). The effects of aerobic exercise training on the age-related lipid peroxidation, Schwann cell apoptosis and ultrastructural changes in the sciatic nerve of rats. *Life Sciences*, *82*(15–16), 840–846.

Skouras, E., Ozsoy, U., Sarikcioglu, L., & Angelov, D. N. (2011). Intrinsic and therapeutic factors determining the recovery of motor function after peripheral nerve transection. *Annals of Anatomy*, *193*(4), 286–303.

Sohnchen, J., Grosheva, M., Kiryakova, S., Hubbers, C. U., Sinis, N., Skouras, E., et al. (2010). Recovery of whisking function after manual stimulation of denervated vibrissal muscles requires brain-derived neurotrophic factor and its receptor tyrosine kinase B. *Neuroscience*, *170*(1), 372–380.

Sterne, G. D., Coulton, G. R., Brown, R. A., Green, C. J., & Terenghi, G. (1997). Neurotrophin-3-enhanced nerve regeneration selectively improves recovery of muscle fibers expressing myosin heavy chains 2b. *The Journal of Cell Biology*, *139*(3), 709–715.

Streppel, M., Azzolin, N., Dohm, S., Guntinas-Lichius, O., Haas, C., Grothe, C., et al. (2002). Focal application of neutralizing antibodies to soluble neurotrophic factors reduces collateral axonal branching after peripheral nerve lesion. *European Journal of Neuroscience*, *15*(8), 1327–1342.

Sullivan, K. A., Kim, B., & Feldman, E. L. (2008). Insulin-like growth factors in the peripheral nervous system. *Endocrinology*, *149*(12), 5963–5971.

Sunderland, I. R., Brenner, M. J., Singham, J., Rickman, S. R., Hunter, D. A., & Mackinnon, S. E. (2004). Effect of tension on nerve regeneration in rat sciatic nerve transection model. *Annals of Plastic Surgery*, *53*(4), 382–387.

Suter, D. M., & Miller, K. E. (2011). The emerging role of forces in axonal elongation. *Progress in Neurobiology*, *94*(2), 91–101.

Tam, S. L., Archibald, V., Jassar, B., Tyreman, N., & Gordon, T. (2001). Increased neuromuscular activity reduces sprouting in partially denervated muscles. *The Journal of Neuroscience*, *21*(2), 654–667.

Teixeira, L. J., Valbuza, J. S., & Prado, G. F. (2011). Physical therapy for Bell's palsy (idiopathic facial paralysis). *Cochrane Database of Systematic Reviews*, *12*, CD006283.

Teng, H. K., Teng, K. K., Lee, R., Wright, S., Tevar, S., Almeida, R. D., et al. (2005). ProBDNF induces neuronal apoptosis via activation of a receptor complex of p75NTR and Sortilin. *The Journal of Neuroscience, 25*(22), 5455–5463.

Topp, K. S., & Boyd, B. S. (2006). Structure and biomechanics of peripheral nerves: Nerve responses to physical stresses and implications for physical therapist practice. *Physical Therapy, 86*(1), 92–109.

Udina, E., Cobianchi, S., Allodi, I., & Navarro, X. (2011). Effects of activity-dependent strategies on regeneration and plasticity after peripheral nerve injuries. *Annals of Anatomy, 193*(4), 347–353.

Udina, E., Puigdemasa, A., & Navarro, X. (2011). Passive and active exercise improve regeneration and muscle reinnervation after peripheral nerve injury in the rat. *Muscle & Nerve, 43*(4), 500–509.

van Meeteren, N. L. U., Brakkee, J. H., Hamers, F. P. T., Helders, P. J. M., & Gispen, W. H. (1997). Exercise training improves functional recovery and motor nerve conduction velocity after sciatic nerve crush lesion in the rat. *Archives of Physical Medicine and Rehabilitation, 78*(1), 70–77.

van Meeteren, N. L., Brakkee, J. H., Helders, P. J., & Gispen, W. H. (1998). The effect of exercise training on functional recovery after sciatic nerve crush in the rat. *Journal of the Peripheral Nervous System, 3*(4), 277–282.

Vaynman, S., Ying, Z., & Gomez-Pinilla, F. (2003). Interplay between brain-derived neurotrophic factor and signal transduction modulators in the regulation of the effects of exercise on synaptic-plasticity. *Neuroscience, 122*(3), 647–657.

Vianney, J.-M., & Spitsbergen, J. M. (2011). Cholinergic neurons regulate secretion of glial cell line-derived neurotrophic factor by skeletal muscle cells in culture. *Brain Research, 1390*, 1–9.

Wehrwein, E. A., Roskelley, E. M., & Spitsbergen, J. M. (2002). GDNF is regulated in an activity-dependent manner in rat skeletal muscle. *Muscle & Nerve, 26*(2), 206–211.

White, C. M., Pritchard, J., & Turner-Stokes, L. (2004). Exercise for people with peripheral neuropathy. *Cochrane Database of Systematic Reviews, 4*, CD003904.

Wilhelm, J. C., Xu, M., Cucoranu, D., Chmielewski, S., Holmes, T., Lau, K., et al. (2012). Cooperative roles of BDNF expression in neurons and Schwann cells are modulated by exercise to facilitate nerve regeneration. *The Journal of Neuroscience, 32*(14), 5002–5009.

Wood, K., Wilhelm, J. C., Sabatier, M. J., Liu, K., Gu, J., & English, A. W. (2011). Sex differences in the effectiveness of treadmill training in enhancing axon regeneration in injured peripheral nerves. *Developmental Neurobiology, 72*(5), 688–698.

Wright, T. W., Glowczewskie, F., Cowin, D., & Wheeler, D. L. (2001). Ulnar nerve excursion and strain at the elbow and wrist associated with upper extremity motion. *The Journal of Hand Surgery, 26*(4), 655–662.

Wright, T. W., Glowczewskie, F., Wheeler, D., Miller, G., & Cowin, D. (1996). Excursion and strain of the median nerve. *The Journal of Bone and Joint Surgery, 78*(12), 1897–1903.

Yamamoto, Y., Livet, J., Pollock, R. A., Garces, A., Arce, V., deLapeyriere, O., et al. (1997). Hepatocyte growth factor (HGF/SF) is a muscle-derived survival factor for a subpopulation of embryonic motoneurons. *Development, 124*(15), 2903–2913.

Yarrow, J. F., White, L. J., McCoy, S. C., & Borst, S. E. (2010). Training augments resistance exercise induced elevation of circulating brain derived neurotrophic factor (BDNF). *Neuroscience Letters, 479*(2), 161–165.

Ying, Z., Roy, R. R., Edgerton, V. R., & Gomez-Pinilla, F. (2003). Voluntary exercise increases neurotrophin-3 and its receptor TrkC in the spinal cord. *Brain Research, 987*(1), 93–99.

Ying, Z., Roy, R. R., Edgerton, V. R., & Gomez-Pinilla, F. (2005). Exercise restores levels of neurotrophins and synaptic plasticity following spinal cord injury. *Experimental Neurology, 193*(2), 411–419.

CHAPTER SEVEN

The Role of Timing in Nerve Reconstruction

Lars B. Dahlin[1]

Department of Clinical Sciences in Malmö/Hand Surgery, Lund University, Skåne University Hospital, Malmö, Sweden
[1]Corresponding author: e-mail address: lars.dahlin@med.lu.se

Contents

Abstract

The surgeon, who treats nerve injuries, should have knowledge about how peripheral nerves react to trauma, particularly an understanding about the extensive pathophysiological alterations that occur both in the peripheral and in the central nervous system. A large number of factors influence the functional outcome, where the surgeon only can affect a few of them. In view of the new knowledge about the delicate intracellular signaling pathways that are rapidly initiated in neurons and in nonneuronal cells with the purpose to induce nerve regeneration, the timing of nerve repair and reconstruction after injury has gained more interest. It is crucial to understand and to utilize the inborn mechanisms for survival and regeneration of neurons and for activation, survival, and proliferation of the Schwann cells and other cells that are acting after a nerve injury. Thus, experimental and clinical data clearly point toward the advantage of early nerve repair and reconstruction of injuries. Following an appropriate diagnosis of a nerve injury, the nerve should be promptly repaired or reconstructed, and new rehabilitation strategies should early be initiated. Considering nerve transfers in the treatment arsenal can shorten the time of nerve reinnervation of muscle targets. Timing of nerve repair and reconstruction is crucial after nerve injury.

A nerve injury has a severe impact on the individual patient, who may experience a broad spectrum of symptom after injury, including sensory dysfunction, lack of muscle function, pain, allodynia, and cold sensitivity. These symptoms, with a profound impact on the patient's global hand and arm function, do not only cause individual suffer to the patient but can also reduce the ability of the patient to enjoy leisure activities and particularly perform their work. Therefore, there is a high risk of long sick leave after a peripheral nerve injury, and the costs for lost production may stand for about 80% of the total cost for treatment (Rosberg et al., 2005; Thorsen, Rosberg, Steen Carlsson, & Dahlin, 2012). The final outcome of a nerve injury depends on many factors. The individual surgeon cannot affect the majority of them, but the time at which a nerve injury should be repaired or reconstructed is important. A patient with a trauma to the hand, arm, shoulder, or lower leg should be properly examined in the emergency room with the attempt to make an appropriate diagnosis of any nerve injury. Once the diagnosis of a nerve injury has been made, all efforts should be directed to repair or reconstruct the nerve injury as soon as possible, although one should consider the general condition of the patient, that is, strict medical priority of the patient's injuries. Based on the delicate intracellular signaling pathways that are rapidly initiated in neurons and non-neuronal cells with the purpose to induce regeneration as well as recent advances in the knowledge about brain plasticity in rehabilitation strategies, the factor *timing* of nerve repair and reconstruction will be the key word.

1. THE INTRINSIC RESPONSE IN NEURONS AND SCHWANN CELLS AFTER INJURY

After a nerve transection, signals are initiated and sent from the site of injury along the proximal segment of the transected axon up to the nerve cell body (Fig. 7.1). These signals may be both positive and negative in nature. The normally transported retrograde signals from targets or from the microenvironment of the axon may be suppressed by the inhibition of axonal transport of substances from the periphery, that is, a negative injury signal. The transcription factor, nuclear factor kappa B, bears a specific code, which is called a nuclear localization sequence, and allows the entrance in the nucleus. Trauma to the nerve may inactivate this factor and thereby it is also an important negative modulator of the response in the nerve cell body (Raivich & Makwana, 2007). After transection, the proximal tip of the axon is rapidly sealed after extracellular cations, such as Na^+ and Ca^{2+}, are diffused through the open cell membrane. The entrance of calcium is also reported to be important for local activation of protein kinases at the tip of the axons. One example of such a locally activated protein kinases is extracellular signal-regulated kinase (ERK) that is fundamental for activation both in neurons and in Schwann cells (Mårtensson, Gustavsson, Dahlin, & Kanje,

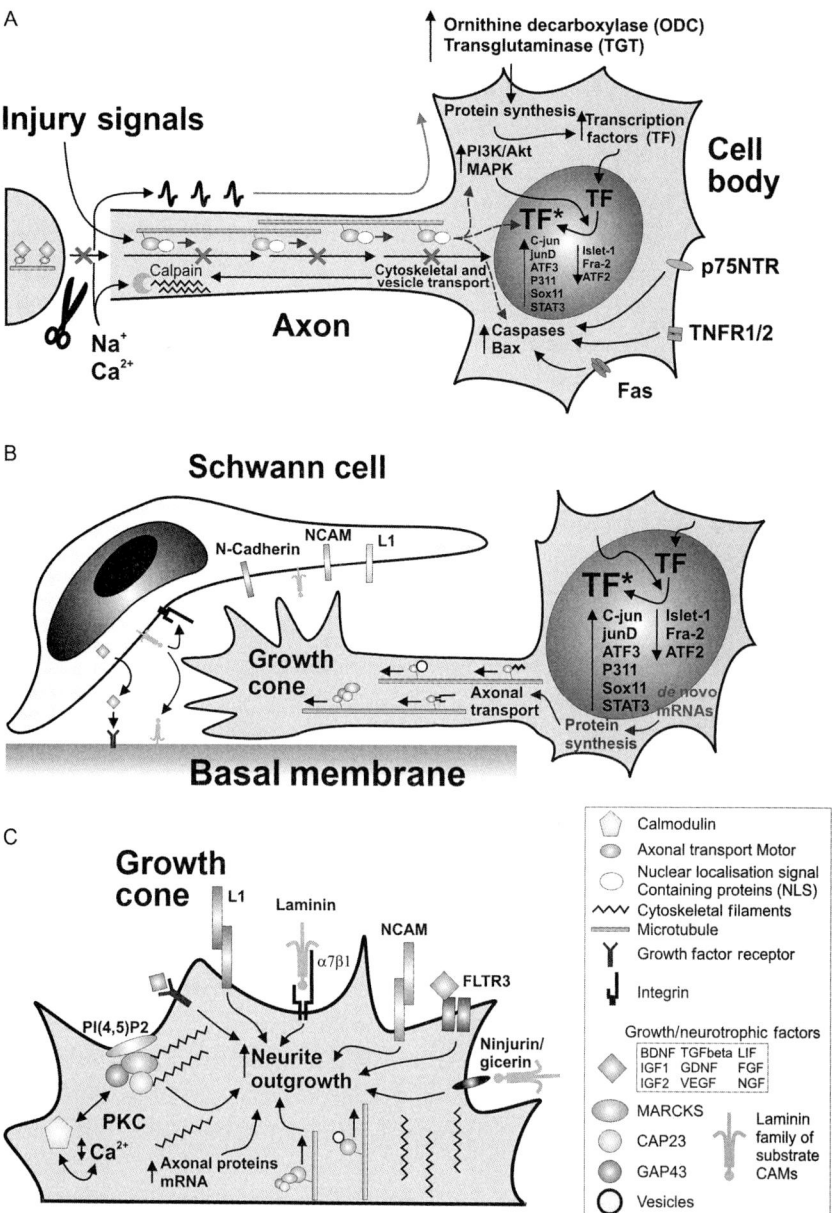

Figure 7.1 Schematic drawing of some of the signals that occur after injury in neurons and in the growth cone at the tip of the axon. After injury, there are both positive and negative injury signals that elicit a number of different events in the nerve cell body and in the nucleus of the neuron; all with the purpose to turn the neuron into a regenerative state instead of transmitting state (A). There is a close contact between the sprout with

(Continued)

2007; Raivich & Makwana, 2007; Stenberg, Kanje, Martensson, & Dahlin, 2011). The microenvironment also contains molecules that are produced by the Schwann cells, by invading inflammatory cells as well as by other injured axons. These substances, like ciliary neurotrophic factor (CNTF) and leukemia inhibitory factor (LIF), can activate the signal transducer and activator of transcription 3 (STAT3) (Wang et al., 2009). Other examples of locally activated molecules are ERK and c-Jun N-terminal kinase (JNK), which are activated by phosphorylation. These complexes are shuttled by retrograde transport with added nuclear localization sequences that allow the translocation to the nucleus. Thereby, these signals are positive injury signals that are generated at or around the tip of the axon with the purpose of translocation to the nucleus for activation of gene transcription (Hanz & Fainzilber, 2004, 2006; Lindwall & Kanje, 2005b) (Fig. 7.1).

The transcription of multiple genes is modulated by retrograde signals, such as the immediate early genes c-Jun and the transcription factors, like activating transcription factor 3 (ATF3) (Hunt, Raivich, & Anderson, 2012); ATF3 being important for a variety of functions (Rynes et al., 2012). These molecules are rapidly upregulated, probably depending on the retrograde transport of JNK, with the purpose of both preserving neuronal survival as well as inducing regeneration (Lindwall, Dahlin, Lundborg, & Kanje, 2004; Lindwall & Kanje, 2005a). STAT3 is also a survival-promoting factor that is present in neurons by using alternative pathways. The activation and compensatory mechanisms are complex, since there is a cross talk between the various signal transection pathways, a phenomenon also seen in Schwann cells (Mårtensson, 2012). Induction of ATF3 may also differ between motor and sensory neurons. The number of neurons that express ATF3 also declines rapidly after an injury, particularly observed in dorsal root ganglia, that is, sensory neurons, which is important when discussing the timing of nerve repair and reconstruction. Thus, an early repair or reconstruction facilitates neuronal survival. In addition, there seems to be a differential upregulation of ATF3 after injury in different types of sensory neurons since the upregulation of ATF3 is particularly important for sensory neurons that project to the skin

Figure 7.1—Cont'd the growth cone and philopodia and the surrounding Schwann cells during the regeneration which occur along important basal membrane that contains laminin and fibronectin (B). In the growth cone actin filaments are assembled depending on local signal transduction mechanisms that stir the direction of the outgrowing axons (C). *Reproduced by kind permission of Raivich and Makwana (2007).* (For color version of this figure, the reader is referred to the online version of this chapter.)

(Reid, Welin, Wiberg, Terenghi, & Novikov, 2010). Thus, a substantial number of different molecules are activated by different mechanisms in neurons after injury. They are all of importance for the survival of neurons and particularly for redirection to the production of substances needed for nerve regeneration (Abe & Cavalli, 2008; Raivich & Makwana, 2007; Rossi, Gianola, & Corvetti, 2007).

The close interaction that normally occurs between axons and Schwann cells is important for maintenance of the differentiation of the Schwann cells. After injury, the Schwann cells are activated (or dedifferentiated) as a preparation before proliferation. In the distal nerve segment, proteases in the axons are rapidly activated for breakdown of the axon and both Schwann cells and the invading and infiltrating macrophages contribute to the breakdown and ingestion of myelin. Myelin-associated genes in Schwann cells are downregulated during this degeneration process. Other molecules, like neural cell adhesion molecule (NCAM), which is linked to non-myelinating Schwann cells, are differently expressed depending on the relation to the outgrowing axons (Saito, Kanje, & Dahlin, 2010). p-ERK1/2 and STAT3 expression is also lower in the distal nerve segment compared to just distal to the site of transection, which may be related to the presence of inflammatory cells or apoptotic Schwann cells. Schwann cells also rapidly express the transcription factor ATF3 as a sign of activation after nerve injury, but such an expression is downregulated in the distal nerve segment in the Schwann cells along with the progress of the outgrowing axons (Kataoka, Kanje, & Dahlin, 2007). Again, these phenomena indicate that there is a close interaction between the outgrowing axons and Schwann cells (Fig. 7.1).

During later phases of the nerve regeneration process, the myelination by ensheathment of the outgrowing axons is promoted by positive regulators between these axons and the different types of Schwann cells, which includes a radial sorting process involving neuregulin-1 (Jessen & Mirsky, 2008). In addition to the role of ERK1/2, recent data also indicate that the transcription factor Pax-3 has a role in differentiation and proliferation of Schwann cells after a peripheral nerve injury (Doddrell et al., 2012). Finally, and maybe of convincing importance, the transcription factor c-Jun, which is expressed in Schwann cells, seems to be a global regulator of the Wallerian degeneration process, since it may determine the expression of trophic factors, adhesion molecules, formation of regeneration tracks, and myelin clearance as well as the control of distinct regenerative potential of the peripheral nerve. If c-Jun is not induced, there is a dysfunction of repair of the cell and with subsequent failure of functional recovery as well as induction of neuronal death. Thus, this single glial transcription factor appears to be of extreme importance in this context (Arthur-Farraj et al., 2012). The c-Jun is also a negative regulator of the myelination (Parkinson et al., 2008). Interestingly, ATF3 expression may be dependent on the c-Jun appearance, particularly observed in sensory neurons (Lindwall et al., 2004).

The knowledge about single transection pathways that are involved in cell death by apoptosis after nerve injury is incompletely known (Mantuano et al., 2011). Apoptosis does occur in some neurons, particularly among sensory neurons, and appears also in satellite cells and in Schwann cells. Those sensory neurons that project to the skin seem to be more vulnerable to apoptosis than those sensory neurons that project to a muscle (Welin, Novikova, Wiberg, Kellerth, & Novikov, 2008). Furthermore, the apoptotic mechanism(s) may be different in Schwann cells and satellite cells compared to motor and sensory neurons, since a marker of apoptosis, cleaved caspase 3, is not observed in neurons after injury, but is clearly expressed in Schwann cells and in some satellite cells (Saito, Kanje, & Dahlin, 2009).

2. THE TIMING OF NERVE REPAIR AND RECONSTRUCTION

As pointed out earlier, there are large numbers of factors that influence the functional outcome after nerve injury and repair and reconstruction; some of them are not possible to influence by the surgeon, like the age of the patient (Chemnitz, Bjorkman, Dahlin, & Rosen, 2013). The timing of nerve repair or reconstruction is an important point in nerve regeneration. It is one of the factors that the surgeon can guide and thereby improve functional outcome. In an interesting clinical paper, it was clearly demonstrated that the results after reconstruction of a brachial plexus nerve injury in adults were considerably better, with respect to motor functional recovery, if the delay of the nerve reconstruction procedure was shorter than 2 weeks (Jivan, Kumar, Wiberg, & Kay, 2009). This clinical observation should be put into perspective with the described mechanisms in signal transduction. An injured neuron with its transected axon can probably be reactivated if a secondary nerve reconstruction procedure is done by "refreshing" the proximal nerve end during surgery through the retransection of the axon. By such a procedure, the described intraneuronal activation mechanisms can probably be reinitiated with a further outgrowth of the axons. However, the obstacle for an efficient axonal outgrowth after delayed nerve repair or reconstruction is the events that occur in the distal nerve segment. The impaired axonal outgrowth after a delayed nerve repair or reconstruction after a transection injury seems to be related to factors like a decrease in p-ERK1/2 in Schwann cells in the distal nerve segment (Tsuda, Kanje, & Dahlin, 2011). Furthermore, the number of Schwann cells that express the transcription factor ATF3, which is also associated with an efficient axonal outgrowth, declines with time after injury. This is particularly obvious in experimental models if the delay exceeds 30 days with a resulting impaired nerve regeneration (Saito & Dahlin, 2008). Apoptosis, detected by the presence of cleaved caspase 3 in Schwann cells, increases substantially in the distal nerve segment if the nerve repair or reconstruction is delayed (Tsuda et al., 2011). Furthermore, the contact between the outgrowing axons and the Schwann cells seems to be essential for the number of cells that express cleaved caspase 3 only if the nerve repair is immediately performed (Tsuda et al., 2011). A delayed nerve repair increases the number of Schwann cells that express cleaved caspase 3 from around 10% of the cells in the distal nerve segment up to around 20% after delayed nerve repair irrespective of the length of the delay (Saito et al., 2009) (Fig. 7.2).

If a nerve injury is not repaired with 3 to 6 months, there is a substantial decline in Schwann cell markers and an increase in fibrosis and proteoglycan scar markers in the distal nerve segment (Jonsson et al., 2013). These changes are similar to those reports that present a decrease in transcription factor, like ATF3, overtime (Saito & Dahlin, 2008). Although it has been suggested that a critical time point at which outcome of regeneration becomes poor appears to be 3 months in experimental systems (Jonsson et al., 2013), even shorter time, like 2 weeks, is sufficient to observe an impaired axonal outgrowth. This phenomenon is related to both a less activation of Schwann cells and an increased number of apoptotic cells (Tsuda et al., 2011).

The cell adhesion molecule NCAM is more abundant in the distal nerve segment if an injured nerve trunk is repaired or reconstructed after a long delay in experimental rat models. After the long delay, when the axonal outgrowth is particularly impaired, NCAM, which is associated with non-myelinating Schwann cells, is predominantly seen in the distal nerve segment. This indicates that particularly nonmyelinating Schwann cells are present in the distal nerve segment if the nerve repair or reconstruction is done with a delay (Saito et al., 2009). Therefore, the myelination of the outgrowing axons may be severely impaired. c-Jun is probably critical in this context, since it reprograms Schwann cells to generate a repair process in the

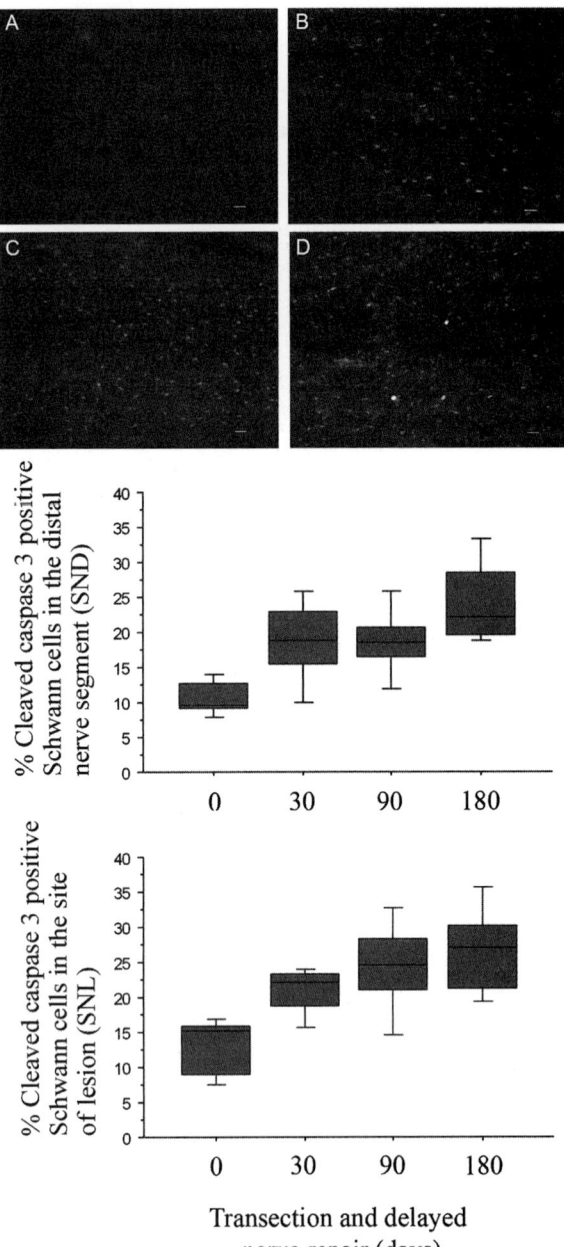

Figure 7.2 Expression of cleaved caspase 3 which is a marker of apoptosis in Schwann cells after transection of a rat sciatic nerve. If a nerve injury is repaired with a delay [30 days delay in (C) and 180 days delayed repair in (D)] compared to an immediate nerve repair (B). The number of Schwann cells that express cleaved caspase 3 increases close to the site of injury (lower diagram) and in the distal nerve segment (upper diagram). Staining of an uninjured nerve is attached in (A) as a comparison. (For color version of this figure, the reader is referred to the online version of this chapter.)

cells after nerve injury (Arthur-Farraj et al., 2012). In addition, the expression of trophic factors and adhesion molecules is determined by c-Jun. Thus, the regeneration tracks, and thereby the regenerative potential of the outgrowing axons, is influenced by c-Jun. The sprouts that are formed from the tip of the axon, with their filopodia, are orchestrated by extracellular guidance cues for their advancement, the retraction, as well as the turning, that is, direction of growth, which is mediated by the actin filaments within the growth cone (Bloom & Morgan, 2011) (Fig. 7.1).

The permissive cues that guide the advancement of the growth cones are critically restored as soon as possible by the surgeon. The previously described basic concepts of nerve repair include preparation of the nerve ends to create a fresh nerve end without necrotic cells and a careful approximation of the nerve ends without tension (Yi & Dahlin, 2010). Compensation for the mushrooming of the fascicles by a meticulous co-aptation, even leaving a minor gap, that allows formation of a fibrin matrix with macrophages and migration of Schwann cells is one of the steps in the surgical procedure. Finally, the nerve repair is completed by application of sutures or glue. These steps are essentially the same when a nerve reconstruction procedure is done, but it is important that the nerve grafts as well as the proximal and distal nerve ends are handled accurately avoiding drying of the tissues. Hereby, the Schwann cells can be kept viable and the signal transduction mechanisms necessary for proliferation and production of growth factors can be preserved.

Different experimental procedures, including pharmacological treatment (Kvist, Danielsen, & Dahlin, 2003), have been described to improve motor reinnervation after a delayed nerve repair (Sulaiman & Gordon, 2009). There is also a possibility to reactivate the chronically denervated Schwann cells by treatment with the transforming growth factor β (TGF-β) (Sulaiman & Gordon, 2009). Recent clinical experiences also indicate that it is a possibility to overcome the problem with chronically denervated Schwann cells by transferring regenerating axons from another source, by a nerve transfer procedure, closer to the target. A variety of nerve transfers have been described for the upper extremity (Lee & Wolfe, 2012). One of these nerve transfer procedures can be done when an axillary nerve injury is overlooked in young adults after shoulder trauma (Dahlin, Coster, Bjorkman, & Backman, 2012). In nerve transfers, a freshly transected nerve fascicle from an uninjured nerve is transferred, without residual problems from the donor nerve, to the distal nerve segment of a previously injured nerve. It is an advantage if the transfer can be made very close to the target.

In this way, the regenerating axons from the freshly transected nerve fascicles need only to grow over a limited distance in an environment with sub-optimal viable Schwann cells before the axons reach a target, such as a dener-vated muscle.

3. CEREBRAL PLASTICITY—THE IMPORTANCE OF TIMING IN REHABILITATION

After a nerve injury, a substantial functional disorganization occurs at the cortical and the subcortical levels in and close to the somatosensory cor-tex in the central nervous system (Rosen et al., 2012; Taylor, Anastakis, & Davis, 2009). These changes are clearly seen in adults, but younger children have an extensive adaptive cerebral plasticity. Therefore, age is an important factor for the functional outcome after nerve injury and repair and recon-struction. There seems to be a critical time around puberty at which the cerebral plasticity after a nerve injury changes and responds as in adults (Chemnitz et al., 2013). Before puberty, the functional recovery may return to almost normal after a nerve injury and repair and reconstruction. The most plausible explanation is the mechanisms of cerebral plasticity, where the functional disorganization of the cortical map can be completely restored in young children. Thus, the brain can in young children more easily inter-pret the new information from the periphery although an extensive misdi-rection of axonal outgrowth occurs (Rosen et al., 2012; Taylor et al., 2009). In addition, there has also been detected a reduction in the cortex after nerve injury. A link between the functional recovery after nerve injury and repair and the disorganization and reorganization of both the gray and white sub-stances has been demonstrated. The described alterations, with a reduced activation of certain brain areas with atrophy of the gray substance as well as the disorganization of reorganization, are rapid. Thus, these changes may also be the target for the factor timing after nerve injury and repair and reconstruction, since they can be considered in the new rehabilitation strategies (Lundborg, Bjorkman, & Rosen, 2007; Weibull et al., 2008). The timing for introduction of training after nerve repair has been highlighted and has focused on the importance of immediate sensory relearning (Rosen & Lundborg, 2007). To provide an alternative afferent inflow from the hand early after nerve repair in the forearm, a "Sensor Glove System" can mediate impulses through the hearing sense. This implies that the depriva-tion of one sense, that is, perception of touch in a denervated skin area, can be compensated by another sense, that is, sensory-by-pass using the "Sensor

Glove System" (Rosen & Lundborg, 2007). By using such a system, initiated early after surgery, with sensory reeducation and compared with the conventional sensory reeducation, which usually starts about 3 months postoperatively, tactile gnosis can significantly be improved (Rosen & Lundborg, 2007). Thus, even the brain can be utilized in the concept of timing after nerve injury and repair and reconstruction (Rosen & Lundborg, 2004).

In conclusion, the functional outcome after nerve injury and repair and reconstruction is dependent on a broad spectrum of different factors and the results cannot be improved by focusing on one of these factors. Therefore, we have to optimize the whole treatment strategies and improve every single part of the whole treatment chain. One of these components is the timing of nerve repair and reconstruction. This component is dependent on that the surgeon makes an early and proper diagnosis of the nerve injury. If a complete nerve injury is suspected, a prompt nerve repair should be done—a statement that is strongly based on recent knowledge about neurobiology. Any delay of the repair or reconstruction leads to a deterioration of the condition of the different cells in the distal nerve segment, particularly the important Schwann cells. The judgment of the extent of injury can be difficult in some specific cases, like in closed nerve injuries and gunshot and shrapnel injuries, where nerve function is impaired, but the continuity of the nerve is preserved. In such cases, one may consider waiting with any tentative surgical procedure of the nerve trunk, but it is of outmost importance that the surgeon in such cases has an attitude of "active surveillance." Thus, the surgeons should evaluate nerve function often and repeatedly. If no signs of functional recovery are observed, the decision of a nerve reconstruction procedure should not be further delayed. The concept of timing does also include an early rehabilitation of the patient, utilizing the new treatment algorithm after nerve repair and reconstruction. Timing of nerve repair and reconstruction after nerve injury is one, but only one in a long chain, key factor to improve functional outcome.

ACKNOWLEDGMENTS

The author's research has been supported by grants from Swedish Research Council (Medicine), Lund University, EU grant (HEALTH-F4-2011-278612 BIOHYBRID), Region Skåne, The Swedish Society for Medicine, Promobilia, HKH Kronprinsessan Lovisa's Fund, and Lundgren's Fund.

This chapter is dedicated to my close friend and research colleague Professor Martin Kanje, who passed away in March 2013. Unfortunately, he could not see this chapter in which he should have been a coauthor, but my work is dedicated to him and his memory. Thank you Martin for all our fantastic and stimulating discussions!

REFERENCES

Abe, N., & Cavalli, V. (2008). Nerve injury signaling. *Current Opinion in Neurobiology, 18*(3), 276–283.

Arthur-Farraj, P. J., Latouche, M., Wilton, D. K., Quintes, S., Chabrol, E., Banerjee, A., et al. (2012). c-Jun reprograms Schwann cells of injured nerves to generate a repair cell essential for regeneration. *Neuron, 75*(4), 633–647.

Bloom, O. E., & Morgan, J. R. (2011). Membrane trafficking events underlying axon repair, growth, and regeneration. *Molecular and Cellular Neurosciences, 48*(4), 339–348.

Chemnitz, A., Bjorkman, A., Dahlin, L. B., & Rosen, B. (2013). Functional outcome thirty years after median and ulnar nerve repair in childhood and adolescence. *The Journal of Bone and Joint Surgery. American Volume, 95*(4), 329–337.

Dahlin, L. B., Coster, M., Bjorkman, A., & Backman, C. (2012). Axillary nerve injury in young adults—An overlooked diagnosis? Early results of nerve reconstruction and nerve transfers. *Journal of Plastic Surgery and Hand Surgery, 46*(3–4), 257–261.

Doddrell, R. D., Dun, X. P., Moate, R. M., Jessen, K. R., Mirsky, R., & Parkinson, D. B. (2012). Regulation of Schwann cell differentiation and proliferation by the Pax-3 transcription factor. *Glia, 60*(9), 1269–1278.

Hanz, S., & Fainzilber, M. (2004). Integration of retrograde axonal and nuclear transport mechanisms in neurons: Implications for therapeutics. *The Neuroscientist, 10*(5), 404–408.

Hanz, S., & Fainzilber, M. (2006). Retrograde signaling in injured nerve—The axon reaction revisited. *Journal of Neurochemistry, 99*(1), 13–19.

Hunt, D., Raivich, G., & Anderson, P. N. (2012). Activating transcription factor 3 and the nervous system. *Frontiers in Molecular Neuroscience, 5*, 7.

Jessen, K. R., & Mirsky, R. (2008). Negative regulation of myelination: Relevance for development, injury, and demyelinating disease. *Glia, 56*(14), 1552–1565.

Jivan, S., Kumar, N., Wiberg, M., & Kay, S. (2009). The influence of pre-surgical delay on functional outcome after reconstruction of brachial plexus injuries. *Journal of Plastic, Reconstructive & Aesthetic Surgery, 62*(4), 472–479.

Jonsson, S., Wiberg, R., McGrath, A. M., Novikov, L. N., Wiberg, M., Novikova, L. N., et al. (2013). Effect of delayed peripheral nerve repair on nerve regeneration, Schwann cell function and target muscle recovery. *PLoS One, 8*(2), e56484.

Kataoka, K., Kanje, M., & Dahlin, L. B. (2007). Induction of activating transcription factor 3 after different sciatic nerve injuries in adult rats. *Scandinavian Journal of Plastic and Reconstructive Surgery and Hand Surgery, 41*(4), 158–166.

Kvist, M., Danielsen, N., & Dahlin, L. B. (2003). Effects of FK506 on regeneration and macrophages in injured rat sciatic nerve. *Journal of the Peripheral Nervous System, 8*(4), 251–259.

Lee, S. K., & Wolfe, S. W. (2012). Nerve transfers for the upper extremity: New horizons in nerve reconstruction. *Journal of the American Academy of Orthopaedic Surgeons, 20*(8), 506–517.

Lindwall, C., Dahlin, L., Lundborg, G., & Kanje, M. (2004). Inhibition of c-Jun phosphorylation reduces axonal outgrowth of adult rat nodose ganglia and dorsal root ganglia sensory neurons. *Molecular and Cellular Neurosciences, 27*(3), 267–279.

Lindwall, C., & Kanje, M. (2005a). Retrograde axonal transport of JNK signaling molecules influence injury induced nuclear changes in p-c-Jun and ATF3 in adult rat sensory neurons. *Molecular and Cellular Neurosciences, 29*(2), 269–282.

Lindwall, C., & Kanje, M. (2005b). The role of p-c-Jun in survival and outgrowth of developing sensory neurons. *Neuroreport, 16*(15), 1655–1659.

Lundborg, G., Bjorkman, A., & Rosen, B. (2007). Enhanced sensory relearning after nerve repair by using repeated forearm anaesthesia: Aspects on time dynamics of treatment. *Acta Neurochirurgica. Supplement, 100*, 121–126.

Mantuano, E., Henry, K., Yamauchi, T., Hiramatsu, N., Yamauchi, K., Orita, S., et al. (2011). The unfolded protein response is a major mechanism by which LRP1 regulates Schwann cell survival after injury. *Journal of Neuroscience, 31*(38), 13376–13385.

Mårtensson, L. (2012). *The role of the MAP- and SAP-kinase pathway in the survival, proliferation and death of Schwann cells of the injured sciatic nerve.* Unpublished Doctoral dissertation, Lund University, Lund.

Mårtensson, L., Gustavsson, P., Dahlin, L. B., & Kanje, M. (2007). Activation of extracellular-signal-regulated kinase-1/2 precedes and is required for injury-induced Schwann cell proliferation. *Neuroreport, 18*(10), 957–961.

Parkinson, D. B., Bhaskaran, A., Arthur-Farraj, P., Noon, L. A., Woodhoo, A., Lloyd, A. C., et al. (2008). c-Jun is a negative regulator of myelination. *Journal of Cell Biology, 181*(4), 625–637.

Raivich, G., & Makwana, M. (2007). The making of successful axonal regeneration: Genes, molecules and signal transduction pathways. *Brain Research Reviews, 53*(2), 287–311.

Reid, A. J., Welin, D., Wiberg, M., Terenghi, G., & Novikov, L. N. (2010). Peripherin and ATF3 genes are differentially regulated in regenerating and non-regenerating primary sensory neurons. *Brain Research, 1310*, 1–7.

Rosberg, H. E., Carlsson, K. S., Hojgard, S., Lindgren, B., Lundborg, G., & Dahlin, L. B. (2005). Injury to the human median and ulnar nerves in the forearm—Analysis of costs for treatment and rehabilitation of 69 patients in southern Sweden. *Journal of Hand Surgery (Edinburgh, Scotland), 30*(1), 35–39.

Rosen, B., Chemnitz, A., Weibull, A., Andersson, G., Dahlin, L. B., & Bjorkman, A. (2012). Cerebral changes after injury to the median nerve: A long-term follow up. *Journal of Plastic Surgery and Hand Surgery, 46*(2), 106–112.

Rosen, B., & Lundborg, G. (2004). Sensory re-education after nerve repair: Aspects of timing. *Handchirurgie, Mikrochirurgie, Plastische Chirurgie, 36*(1), 8–12.

Rosen, B., & Lundborg, G. (2007). Enhanced sensory recovery after median nerve repair using cortical audio-tactile interaction. A randomised multicentre study. *Journal of Hand Surgery. European Volume, 32*(1), 31–37.

Rossi, F., Gianola, S., & Corvetti, L. (2007). Regulation of intrinsic neuronal properties for axon growth and regeneration. *Progress in Neurobiology, 81*(1), 1–28.

Rynes, J., Donohoe, C. D., Frommolt, P., Brodesser, S., Jindra, M., & Uhlirova, M. (2012). Activating transcription factor 3 regulates immune and metabolic homeostasis. *Molecular and Cellular Biology, 32*(19), 3949–3962.

Saito, H., & Dahlin, L. B. (2008). Expression of ATF3 and axonal outgrowth are impaired after delayed nerve repair. *BMC Neuroscience, 9*, 88.

Saito, H., Kanje, M., & Dahlin, L. B. (2009). Delayed nerve repair increases number of caspase 3 stained Schwann cells. *Neuroscience Letters, 456*(1), 30–33.

Saito, H., Kanje, M., & Dahlin, L. B. (2010). Crossed over repair of the femoral sensory and motor branches influences N-CAM. *Neuroreport, 21*(12), 841–845.

Stenberg, L., Kanje, M., Martensson, L., & Dahlin, L. B. (2011). Injury-induced activation of ERK 1/2 in the sciatic nerve of healthy and diabetic rats. *Neuroreport, 22*(2), 73–77.

Sulaiman, O. A., & Gordon, T. (2009). Role of chronic Schwann cell denervation in poor functional recovery after nerve injuries and experimental strategies to combat it. *Neurosurgery, 65*(4 Suppl.), A105–A114.

Taylor, K. S., Anastakis, D. J., & Davis, K. D. (2009). Cutting your nerve changes your brain. *Brain, 132*(11), 3122–3133.

Thorsen, F., Rosberg, H. E., Steen Carlsson, K., & Dahlin, L. B. (2012). Digital nerve injuries: Epidemiology, results, costs, and impact on daily life. *Journal of Plastic Surgery and Hand Surgery, 46*(3–4), 184–190.

Tsuda, Y., Kanje, M., & Dahlin, L. B. (2011). Axonal outgrowth is associated with increased ERK 1/2 activation but decreased caspase 3 linked cell death in Schwann cells after immediate nerve repair in rats. *BMC Neuroscience, 12*, 12.

Wang, L., Lee, H. K., Seo, I. A., Shin, Y. K., Lee, K. Y., & Park, H. T. (2009). Cell type-specific STAT3 activation by gp130-related cytokines in the peripheral nerves. *Neuroreport, 20*(7), 663–668.

Weibull, A., Bjorkman, A., Hall, H., Rosen, B., Lundborg, G., & Svensson, J. (2008). Opti-
mizing the mapping of finger areas in primary somatosensory cortex using functional
MRI. *Magnetic Resonance Imaging, 26*(10), 1342–1351.

Welin, D., Novikova, L. N., Wiberg, M., Kellerth, J. O., & Novikov, L. N. (2008). Survival
and regeneration of cutaneous and muscular afferent neurons after peripheral nerve
injury in adult rats. *Experimental Brain Research, 186*(2), 315–323.

Yi, C., & Dahlin, L. B. (2010). Impaired nerve regeneration and Schwann cell activation after
repair with tension. *Neuroreport, 21*(14), 958–962.

CHAPTER EIGHT

Future Perspectives in Nerve Repair and Regeneration

Pierluigi Tos[*,1], **Giulia Ronchi**[†,1], **Stefano Geuna**[†,2], **Bruno Battiston**[*]

[*]Department of Traumatology, Microsurgery Unit, C.T.O. Hospital, Città della Scienza e della Salute, Turin, Italy
[†]Department of Clinical and Biological Sciences, Neuroscience Institute of the Cavalieri Ottolenghi Foundation (NICO) University of Turin, Orbassano, Italy
[1]These authors contributed equally to this work
[2]Corresponding author: e-mail address: stefano.geuna@unito.it

Contents

Abstract

After peripheral nerve injuries, the process of nerve regeneration and target rein-nervation is very complex and depends on many different events occurring not only at the lesion site but also proximally and distally to it. In spite of the recent scientific and technological advancements, the need to find out new strategies to improve clinical nerve repair and regeneration remains. To reach this goal, the therapeutic strategy should thus exert its effects at different levels in order to simultaneously potentiate axonal regeneration, increase neuronal survival, modulate central reorganization, and inhibit or reduce target organ atrophy. It is expected that this multilevel approach might lead to significant improvement in the functional outcome and thus the quality of life of the patients suffering from peripheral nerve injury.

International Review of Neurobiology, Volume 109
ISSN 0074-7742
http://dx.doi.org/10.1016/B978-0-12-420045-6.00008-0
165

1. INTRODUCTION

Peripheral nerve injuries belong to the most challenging and difficult surgical reconstructive problems, and often cause partial or total loss of motor, sensory, and autonomic functions. The consequences of nerves injuries, which occur in approximately 2.8% of trauma patients (Huelsenbeck et al., 2012), may be disastrous and can result in substantial functional loss, thus interfering with many aspects of a person's life because of permanently impaired sensory and motor functions. Moreover, development of secondary problems, such as neuropathic pain, dysesthesia, and cold intolerance is frequently observed following nerve injuries. In addition, nerve injuries have also a substantial economic impact on the society in terms of health care and long periods of sick leave (de Putter et al., 2012).

Despite the ability of the peripheral nerve to regenerate and reinnervate denervated target organs has been recognized for more than a century, clinical and experimental evidences show that the regeneration is usually far from satisfactory, especially after severe injuries (Navarro, Vivo, & Valero-Cabre, 2007; Pfister et al., 2011; Sun et al., 2009). So far, there is no technique to guarantee total recovery and normalization of functional sensibility following repair of an injured nerve. The poor outcome reflects the complexity of peripheral nerve injuries and the diversity of cellular and biochemical events, which are required to regain function. Indeed, a nerve injury differs from most other types of tissue injuries in the body, since it is not only a local repair process that is required. The processes of nerve regeneration and target reinnervation are complex, involving many factors which lead to immediate as well as long-term physiological, biochemical, and cellular changes (Fig. 8.1) (Lundborg, 2005).

First of all, dramatic changes occur at the level of the damaged nerve (Geuna et al., 2009). After a peripheral nerve traumatic lesion, at the level of the nerve injury, changes begin almost immediately, both proximally and distally to the lesion. In the proximal segment, axons degenerate for some distance back from the site of injury. Within hours after injury, the axon produces a great number of collateral sprouts that advance distally. After nerve transection, the distal segment undergoes a slow process of degeneration known as Wallerian degeneration, which starts immediately after injury and involves myelin breakdown and proliferation of Schwann cells. Schwann cells and macrophages are recruited to the injury site and phagocytize all the myelin and cellular debris. As the axon sprouts from the

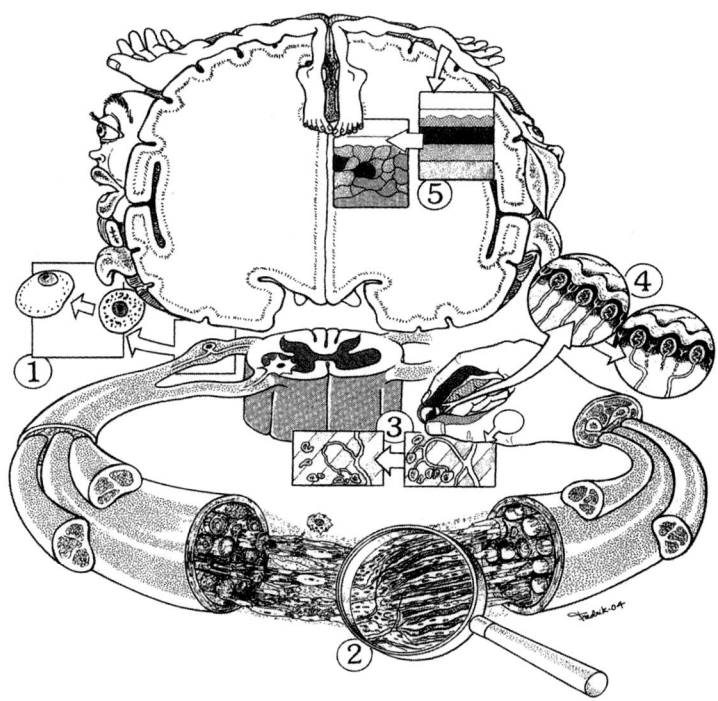

Figure 8.1 Events occurring at various levels as a response to nerve injury and repair. (1) In the cell bodies in dorsal root ganglia and anterior horns of the spinal cord; (2) at the site of injury; (3,4) at target level—motor endplates and cutaneous innervation; and (5) at the cortical level (Lundborg, 2005).

proximal stump, they regenerate between the layers of basal lamina of the Schwann cell processes reaching finally the target organs (Geuna et al., 2009).

Second, changes occur to the target organ innervated by the damaged nerve (Geuna et al., 2009). The regenerated axon must reinnervate the proper target, and the target must retain the ability to accept reinnervation and recover from denervation-related atrophy. Regeneration rate is approximately 1/2 mm/day; therefore, more proximal injuries lead to longer denervation periods and, despite optimal microsurgical techniques, the functional results achieved after repair of severed peripheral nerves are much less than optimal due to the target organ atrophy that takes place during the period of denervation.

Third, changes occur to the proximal neural structures (e.g., dorsal root ganglia and spinal cord) where the cell bodies of the neurons are located. As a consequence of peripheral nerve injury, cell bodies in dorsal root ganglia (DRGs) and anterior horns of the spinal cord undergo adaptive changes that involve a chromatolytic reaction associated with a shift in protein synthesis from a "signaling mode" to a "growing mode" and protein synthesis switches from neurotransmitter-related substances to those required for axonal reconstruction. Moreover, the peripheral and central nervous systems (CNSs) are functionally integrated and a peripheral nerve lesion always results in long-lasting central modifications and reorganization (Kaas, 1991; Kaas & Collins, 2003; Wall, Xu, & Wang, 2002). The mechanisms of plasticity and reorganization of brain circuits that occur after nerve injury are complex; they may result in beneficial adaptive functional changes or contrarily cause maladaptive changes, such as pain, dysesthesia, hyperreflexia, and dystonia (Lundborg, 2000, 2003).

2. CHANGES AT THE NERVE LEVEL

In 1943, Sir Herbert Seddon introduced a classification of three discrete types of nerve injury: neurapraxia, axonotmesis, and neurotmesis (Seddon, 1943):

i. Neurapraxia is a mild injury characterized by local myelin damage. Axon continuity is preserved, and the nerve does not undergo Wallerian degeneration. It may result from exposure to a wide range of conditions such as heat, cold, irradiation, or electrical injuries, but is most commonly due to mechanical stress, such as concussion, compression, or traction injuries. Recovery may occur within hours, days, weeks, or up to a few months.

ii. Axonotmesis involves additional damage to peripheral axons, but connective tissue structures remain intact. The interruption of axons is often the result of nerve pinching, crushing, or prolonged pressure. Wallerian degeneration occurs, but subsequent axonal regrowth may proceed along the intact endoneurial tubes. Recovery depends upon the degree of internal disorganization in the nerve as well as the distance to the end organ.

iii. Neurotmesis is the most severe injury, equivalent to physiologic disruption of the entire nerve. Functional recovery does not easily occur because of the extent of endoneurial tube disruption. Nonetheless, successful regeneration might result with surgical intervention.

In 1951, Sunderland expanded Seddon's classification to five degrees of peripheral nerve injury instead of three (Sunderland, 1951). He divided

Seddon's axonotmesis grade into three types, depending on the degree of connective tissue involvement:

i. Type 1 injury corresponds to Seddon's neurapraxia with conduction block and completely intact stroma.

ii. Type 2 injury corresponds to Seddon's axonotmesis. The endoneurium, perineurium, and epineurium are still intact, but the axons are physiologically disrupted. Recovery can occur by axonal regrowth along endoneurial tubes, and complete functional recovery can be expected. The time for recovery depends on the level of injury, usually months.

iii. In type 3 injury, the endoneurium is also disrupted, but the surrounding perineurium and epineurium are intact. Recovery is incomplete and depends upon how well the axons can cross the site of the lesion and find endoneurial tubes.

iv. In type 4 injury, individual nerve fascicles are transected, and the continuity of the nerve trunk is maintained only by the surrounding epineurium. This type of injury requires surgical repair or reconstruction of the nerve.

v. Type 5 injury is equivalent to Seddon's neurotmesis (complete nerve disruption), and spontaneous recovery is negligible.

Although Sunderland's classification provides a concise and anatomic description of nerve injury, the clinical utility of this system is debatable since a nerve may undergo a combination of different degrees of injury. Therefore, a sixth degree of nerve injury has been introduced to define a combination of the other degrees of injuries (Mackinnon, 1989).

After a peripheral nerve traumatic injury, complex pathophysiologic changes, including morphologic and metabolic changes, occur at the injury site almost immediately. The interruption of a peripheral nerve causes significant changes in normal morphology and tissue organization both proximally and distally to the lesion site.

The nerve stump distal to the lesion undergoes a degeneration that is now known as "Wallerian degeneration" in honor of Augustus Volney Waller, who first characterized the disintegration of the frog glossopharyngeal and hypoglossal nerves after axotomy 160 years ago (Stoll, Jander, & Myers, 2002). The process involves a number of phases, some concurrent, others consecutive, in which the distal portions of all affected axons degenerate. The sequence begins with prompt degradation of axoplasm, axolemma, and myelin sheath due to proteolysis (Lubinska, 1982; Schlaepfer, 1977; Vial, 1958). Then, the degraded myelin is phagocytized by the recruited macrophages to aid the removal of axonal and myelin debris (Bruck, 1997; Vargas & Barres, 2007). Fragmentation of axons is first detected few hours after nerve transection. Within 48 h, the entire nerve is fully involved, and over a period of 3–6 weeks, Schwann cells and macrophages phagocytize all the myelin and cellular debris.

Schwann cells play several important roles in nerve degeneration and regeneration:

i. Coincident with axonal injury, Schwann cells in the distal nerve begin to dedifferentiate (Lee et al., 2009). Within 48 h of injury, they start altering their gene expression: expression of myelin proteins (e.g., P0, MAG (myelin-associated glycoprotein)) (Trapp, Hauer, & Lemke, 1988; White et al., 1989) and connexin 32 (a gap junction protein which forms reflexive contacts within individual myelinating SCs at para-nodes) (Hall, 2001) decreases dramatically as a consequence of axonal degeneration distal to the injury site, whereas regeneration-associated genes (GAP-43), neurotrophic factors and their receptors, neurotrophin 4/5 (NT-4/5) neuregulin and its receptors, including the low-affinity neurotrophin receptor p75NTR; nerve growth factor (NGF), brain-derived neurotrophic factor (BDNF), glial cell line-derived neurotrophic factor (GDNF), and insulin-like growth factors (IGFs) (Carroll, Miller, Frohnert, Kim, & Corbett, 1997; Hall, 2001) are upregulated.

ii. Between days 1 and 5 after injury, Schwann cells start proliferating, a critical event for the promotion of axonal regeneration. Their peak of activation occurs around day 3 postinjury and then decreases during the following weeks. A second phase of proliferation occurs during the regenerative process. Proliferating Schwann cells align in columns known as bands of Büngner, which provide a supportive substrate and growth factors for regenerating axons (Griffin & Thompson, 2008; Stoll, Griffin, Li, & Trapp, 1989).

iii. Schwann cells also play an important role in removing myelin debris; rapid clearance of myelin appears to be the most important precondition for axonal regeneration after peripheral nerve injury because it contains molecules that are inhibitory to axonal growth, such as MAG and oligodendrocyte-myelin glycoprotein (Huang et al., 2005; Shen et al., 2000).

iv. Yet, Schwann cells in the distal nerve stump produce several neuro-trophic and neurotropic molecules (such as NGF, BDNF, NT-4, GDNF, and insulin-like growth factor-1 (IGF-1)) that promote axon growth (Chen, Yu, & Strickland, 2007). Moreover, denervated Schwann cells overexpress fibronectin, laminin, tenascin, and some proteoglycans, which create a substrate for axonal elongation. The phenotype of reactive Schwann cells resembles the one of immature Schwann cells, and they form a permissive substrate for regeneration. When these cells regain contact with the axons, they redifferentiate again (Jessen & Mirsky, 2008).

Wallerian degeneration represents the basis for the nerve regeneration and target reinnervation processes (Battiston, Geuna, Ferrero, & Tos, 2005; Schmidt & Leach, 2003; Terzis, Sun, & Thanos, 1997). In severe injuries, nerve regeneration begins only after Wallerian degeneration has run its course, but in mild injuries, the regenerative and repair processes begin almost immediately. The rate of axonal regrowth is determined by changes within the cell body, the activity of the specialized growth cone at the tip of each axon sprout, and the resistance of the injured tissue between cell body and end organ.

Regenerating axons are usually produced at the node of Ranvier located close to the proximal stump of the lesion (Hopkins & Slack, 1981; McQuarrie, 1985). Nodal sprouts usually contain vesicles of various sizes, and as the sprouts develop into well-formed growth cones, the number of vesicles markedly increases. Sprouts from the node of Ranvier extend through their own basal lamina tubes in the proximal segment, traverse the narrow gap of connective tissue between the proximal and distal stumps, and finally enter the distal nerve segment.

During extension through the Schwann cell column, regenerating axons grow along the Schwann cell basal lamina. Axon–Schwann cell attachment is mediated by various adhesion molecules including the immunoglobulin superfamily, for example, neural cell adhesion molecule (N-CAM) and L1, and the cadherin superfamily, for example, N-cadherin and E-cadherin, whereas axon basal lamina contact is for the most part mediated by laminin (Letourneau, Condic, & Snow, 1994). These adhesion molecules are no longer detected when Schwann cells begin to form the myelin sheath around the axon, whereas the mature unmyelinated fibers continue to exhibit such adhesion molecules (Ide, 1996).

When surgical repair of the nerve is required, the goal is to guide regenerating sensory, motor, and autonomic axons to the distal nerve segment to maximize the chance of target reinnervation (Pfister et al., 2011). Nerve reconstruction by tissue engineering has seen an increasing interest over the past years (Leach & Schmidt, 2005; Pfister et al., 2011). Despite the spontaneous regeneration potential of peripheral nerves and the best efforts and modern surgical techniques, functional restoration is often incomplete and clinical results are still unsatisfactory (Battiston et al., 2009; Scholz et al., 2009).

Peripheral nerve injury may result in injury without gaps or injury with gaps between the nerve stumps. When there is no gap or the gap is short (~5 mm or less), as in simple injuries, the common surgical approach is a direct suture of the two stumps (end-to-end suture) (Terzis et al., 1997).

For longer nerve gaps, when nerve injury resulted in substance loss between the two nerve stumps, this direct suturing under tension leads to very poor clinical results (Dvali & Mackinnon, 2003) and a segment of nerve or other materials must be used to bridge the gap.

The demonstration, in the early 1970s, that grafting of an autogenous nerve segment to bridge a nerve defect leads to better clinical results than suturing the two stumps under tension (Millesi, 1970), opened a new era in peripheral nerve surgery making it possibly the most successful surgical approach to complex lesions that before would have been unfathomable.

There are three types of conventional bridging materials:

i. *Autologous nerve grafts.* Nerve autografts have seen extensive clinical employment over the past 30 years. A nerve graft provides an ideal conduit for regenerating axons because it provides a scaffold which contains Schwann cell basal laminae, and moreover, these Schwann cells produce growth factors (Lundborg, 2004). Autogenous nerve grafting can be performed with nonvascularized autogenous nerve, vascularized nerve, interpositional conduits, and nerve allografts. However, it has several disadvantages, including an extra incision for the removal of a healthy sensory nerve, and the removal of a healthy sensory nerve which will result in a sensory deficit. Finally, donor graft material is limited, particularly for managing extensive lesions which require several lengths of nerve graft.

ii. *Non-nervous biological grafts.* Conduits made by small segments of an artery were first successfully employed by Bungner (Bungner, 1891). However, interest shifted then to veins for their larger availability and reduced side effects related to their withdrawal (Wrede, 1909). Similar to veins, also the use of skeletal muscle autografts for nerve repair was already reported many years ago (Fawcett & Keynes, 1990; Keynes, Hopkins, & Huang, 1984). The idea of employing muscle fibers for axonal regeneration is on the similarities between the muscle basal lamina and the endoneurial tubes (Fawcett & Keynes, 1990;

Glasby, Gschmeissner, Hitchcock, & Huang, 1986). Finally, a combined conduit by enriching vein segments with fresh skeletal muscle fibers (muscle-in-vein conduit) is used to improve effectiveness of tubulization nerve repair (Battiston, Tos, Cushway, & Geuna, 2000; Battiston, Tos, Geuna, Giacobini-Robecchi, & Guglielmone, 2000; Brunelli, Battiston, Vigasio, Brunelli, & Marocolo, 1993; Fornaro, Tos, Geuna, Giacobini-Robecchi, & Battiston, 2001; Tos, Battiston, Ciclamini, Geuna, & Artiaco, 2012).

iii. *Non-biological grafts.* The use of non-biological materials for nerve reconstruction has a lengthy history started at the beginning of the twentieth century, and many attempts to use various nonbiological materials, such as metals, permeable cellulose esters, gelatine tubes, rubber, plastics, etc., were carried out (Fields, Le Beau, Longo, & Ellisman, 1989). The past 30 years saw an impressive increase of experimental studies aimed at testing new biomaterials for nerve regeneration, such as decalcified silicone tube, bone tube, nylon fiber tube, polyurethanes, etc. (Battiston et al., 2005; Pfister, Papaloizos, Merkle, & Gander, 2007; Schmidt & Leach, 2003). The results have been in general very successful, and their effectiveness is similar and sometimes even superior to traditional nerve autografts (Navarro et al., 1996; Yannas & Hill, 2004; Young, Wiberg, & Terenghi, 2002).

Moreover, direct neurotization of denervated muscles is used in situations where the motor nerve has been avulsed and direct nerve suture or grafting is not possible (Brunelli, 2005). It has been demonstrated that an axon that is in contact with a denervated muscular fiber can form a new neuromuscular junction (NMJ). A prerequisite for this procedure is that there is some residual trophism of the muscle. Generally, however, neurotization procedures have poor functional outcome.

Finally, end-to-side neurorrhaphy is based on the assumption that an intact nerve can "donate" axons to the distal end of an injured nerve (Papalia et al., 2003). This technique has received particular interest when the nerve gap is large or when the lesion is proximal, both of which severely limit nerve regeneration.

3. CHANGES OCCURRING DISTALLY TO THE DAMAGED NERVE: FOCUS ON SKELETAL MUSCLE

Target organ atrophy might represent a limiting factor in functional recovery after nerve repair and regeneration. Among the different sensory and motor target organs, skeletal muscles represent the most important ones in terms of clinical relevance. The normal structural and functional integrity of skeletal muscle depends on intact innervation, normal transmission of impulses across the myoneural junction, and normal metabolic processes within the muscle cell. Injury to peripheral nerves always results in immediate loss of muscle function and progressive skeletal muscle atrophy, thus representing an important cause of poor clinical results after nerve reconstruction. Following a peripheral nerve injury, the longer the interval between denervation and reinnervation, the poorer the degree of motor recovery; thus, the regenerative outcome may be very poor when reinnervation of denervated target organs is delayed due either to a long distance between target and lesion site or to delayed nerve repair following major trauma (Birch & Raji, 1991; Merle, Bour, Foucher, & Saint Laurent, 1986). Since axons usually regrow at an average rate of 1 mm/day (Buchthal & Kuhl, 1979; Seddon & Fynn, 1972), it would take a long time for the muscle to be reinnervated. The success of reinnervation depends therefore both on the ability of the neuron to reprogram its growth and to establish new connections, and on the ability of muscle fibers to survive in the absence of trophic and regulating signals derived from the nerve.

Denervated muscles come across structural, biochemical, and physiological changes eventually leading to atrophy and apoptosis, losing up to 80% of their mass (Gutmann, 1962). Over time, denervated muscles lose receptiveness to regenerated motor axons that reach the muscle because of a significant loss of viable muscle cells due to fiber necrosis, connective tissue hyperplasia, and exhaustion of satellite cell regeneration (Fu & Gordon, 1995; Irintchev, Draguhn, & Wernig, 1990; Schmalbruch, al-Amood, & Lewis, 1991; Veltri, Kwiecien, Minet, Fahnestock, & Bain, 2005). The loss of neural input, including neurotransmitters, neurotrophic factors, and other signals, promotes muscle fiber atrophy and thus reduces receptivity to regenerated axons (Veltri et al., 2005). Moreover, in response to an injury, satellite cells undergo a period of rapid proliferation; the majority of the satellite cells differentiate and fuse to form new myofibers or to repair the damaged ones (Lu, Huang, & Carlson, 1997; Schultz, Jaryszak, & Valliere, 1985).

During early stages, denervated muscle shows also a wide spectrum of molecular and cellular changes, including changes in gene expression. Upon denervation, there is the upregulation of NGF, BDNF (Zhao, Veltri, Li, Bain, & Fahnestock, 2004), IGF (Tang, Cheung, Ip, & Ip, 2000), fibroblast growth factor (FGF), hepatocyte growth factor (Yamaguchi, Ishii, Morita, Oota, & Takeda, 2004), and the alpha component of ciliary neurotrophic factor (CNTF) receptor complex (Tang et al., 2000). Also, the expression of many metabolic molecules, such as ferritin heavy chain, adhesion molecules, such as N–CAM, and extracellular proteases, including urokinase, change in response to muscle denervation. Finally, an increase in the level of ErbB2 and ErbB3 receptors and Neuregulin1 expression was also demonstrated (Ng, Pun, Yang, Ip, & Tsim, 1997; Nicolino et al., 2009; Suarez et al., 2001). These adaptive changes might act to maintain muscle fiber survival during early stages of denervation and participate in the remodeling of neuromuscular synapse (Tang et al., 2000).

Traditional strategies to improve motor functional recovery after injury by delaying the effects of the denervation process include electrical stimulation and rehabilitation of the denervated muscles (Nicolaidis & Williams, 2001). These treatments can improve muscle function after nerve injury in the clinical setting; however, they are not very effective in arresting denervated muscle atrophy and patient compliance is often poor; moreover, implantable electrical systems are expensive.

Microsurgical repair within 2 months of injury can essentially reverse skeletal muscle changes and result in good functional recovery (Finkelstein, Dooley, & Luff, 1993). In contrast, if surgery is delayed for 6 months or more, denervation results in irreversible structural damage, including extrafusal fiber necrosis, connective tissue hyperplasia, and deterioration of the muscle spindles, leading to poor reinnervation and functional recovery (Bain, Veltri, Chamberlain, & Fahnestock, 2001; Hynes, Bain, Thoma, Veltri, & Maguire, 1997; Veltri et al., 2005). Furthermore, even when nerve surgery is performed early, there will still be a long period of muscle denervation if the distance from the site of injury is substantial, and the operative results are likely to be correspondingly poor.

A useful strategy to delay the skeletal muscle atrophy might be to connect the end of a sensory nerve to the side of the distal nerve stump of the injured nerve (sensory protection) in order to maintain the structural and functional integrity of muscle until axons of the native nerve reach their target (Bain et al., 2001; Hynes et al., 1997; Irintchev et al., 1990; Veltri et al., 2005; Wang, Gu, Xu, Shen, & Li, 2001). This strategy uses a readily available sensory nerve to directly or indirectly support denervated muscle fibers by the supply of trophic factors, improve existing endoneurial sheath structure, and enhance regeneration by the native nerve (Veltri et al., 2005; Zhao et al., 2004). It has been shown that sensory protection minimizes two of the three major structural consequences of chronic denervation: fiber necrosis and connective tissue hyperplasia.

Another solution to delay the denervated atrophy of skeletal muscles is the use of neural stem cells (NSCs); recent studies have reported that genetically modified NSCs ameliorate experimental spinal muscular atrophy, providing neurotrophic support or partially replacing interrupted innervation between neural cells and skeletal muscles (Corti et al., 2006, 2009). Embryonic stem cells transplanted into the spinal cord could differentiate into relatively normal motor neurons, extend axons into periphery nerves, and form new NMJs with denervated muscles (Deshpande et al., 2006; Gao, Coggeshall, Tarasenko, & Wu, 2005; Lee, Tos, et al., 2007; Lee, Jeyakumar, et al., 2007).

Finally, local and/or systemic administration of various molecules might also prevent skeletal muscle atrophy, such as IGF-1 (Latres et al., 2005; Yoshida, Semprun-Prieto, Sukhanov, & Delafontaine, 2010) and ghrelin (Porporato et al., 2013).

4. CHANGES OCCURRING PROXIMALLY TO THE DAMAGED NERVE

4.1. Changes in the DRGs after peripheral nerve injury and regeneration

Anatomically, the DRGs are located along the dorsal spinal roots; they house the cell bodies of primary afferents of the spinal sensory system and are surrounded by a thick connective capsule. As a consequence of a peripheral nerve injury, trophic support from the periphery is blocked and DRG neuron cell bodies undergo adaptive changes: Nissl bodies (i.e., the basophilic neurotransmitter synthetic machinery) undergo dissolution, which is followed by a prominent migration of the nucleus from the center of the cell toward the periphery, an increase in the size of the nucleolus, nucleus, and cell body, cell swelling, and retraction of dendrites, which collectively are called "chromatolytic changes" (Fenrich & Gordon, 2004; Lieberman, 1971). The disappearance of the prominent basophilic-stained Nissl granules is particularly evident. These granules are ribosome clusters of rough endoplasmic reticulum, that is not observed after axotomy, when they become disorganized, freeing polyribosomes and ribonucleotides into the cytoplasm. The severity and the time course of the chromatolytic process are mainly influenced by the severity of the injury, the distance of lesion to cell body, the type of neuron, and the age (Navarro et al., 2007).

The dissolution of the ribosomes and ordered arrays of rough endoplasmic reticulum that constitute the Nissl bodies are then accompanied by metabolic changes including overall increases in protein and mRNA synthesis as well as changes in the pattern genes that the neuron expresses. The main metabolic activity of the cell is shifted from synthesizing neurotransmitter-related proteins to the synthesis of structural materials needed for axon repair and growth. For example, choline acetyltransferase is downregulated, whereas the neuropeptide, calcitonin gene-related peptide, the fast transported growth-associated protein, GAP-43, and the slowly transported cytoskeletal proteins, actin and tubulin, are upregulated (Haas, Donath, & Kreutzberg, 1993; Tetzlaff, Gilad, Leonard, Bisby, & Gilad, 1988). Glucose-6-phosphate dehydrogenase and hydrolytic enzyme are also upregulated (Davis, Taylor, & Anastakis, 2011; Fawcett & Keynes, 1990).

The success of nerve regeneration and functional reinnervation of targets first depend on the capacity of axotomized neurons to survive and shift toward the regenerative phenotype (Navarro et al., 2007). Results of studies on the changes in DRG neuron number following a nerve injury show a wide variation in results. Most of the authors report that peripheral nerve transection induces the primary sensory neuronal death (Himes & Tessler, 1989; Terenghi, 1999; Vestergaard, Tandrup, & Jakobsen, 1997), showing that between 7% and 50% of primary sensory neurons (more small than large neurons) die after injury (Himes & Tessler, 1989; McKay Hart, Brannstrom, Wiberg, & Terenghi, 2002). Other authors report no significant neuron loss (Swett, Hong, & Miller, 1995) or no detectable loss of dorsal root axons until 4 months (Coggeshall, Lekan, Doubell, Allchorne, & Woolf, 1997; Cohen, Yachnis, Arai, Davis, & Scherer, 1992) after injury to the spinal or sciatic nerve.

Further, during regeneration, the cell body undergoes visible changes that mark the reversal of chromatolysis. Indeed, the nucleus returns to the cell center and nucleoproteins reorganize into the compact Nissl granules. A complex and incompletely understood interaction occurs between the cell body and the regenerating axon tip. Axoplasm, which serves to regenerate the axon tip, arises from the proximal axon segment and cell body. Both fast and slow components of axoplasmic transport supply materials from the cell body to the sites of axonal regeneration. The rate of increase in protein and lipid synthesis in the cell body influences the rate of advance and the final caliber of the regenerating axon (Burnett & Zager, 2004).

4.2. CNS plasticity induced by peripheral nerve injury and regeneration

Until recently, it was thought that no new neural connections could be formed in the adult brain (Kandel & Squire, 2000). It was assumed that, once connections had been established in foetal life, they hardly changed in adulthood. The only areas of the adult brain capable of reorganization were those involved in learning and memory processes.

The picture has changed radically in the past decades; indeed, recent evidence demonstrates that the brain is capable of remarkable and widespread adaptive changes in response to peripheral injuries (Davis et al., 2011; Jain, Florence, & Kaas, 1998; Navarro et al., 2007). In fact, an injured nerve stops to function normally and this occurrence results in the reorganization of the projections to the CNS. While it is likely that this reorganization following injury takes place in the cortex, plastic changes may also occur in subcortical structures such as the thalamus, brainstem relay nuclei, and spinal cord (Lewin & McMahon, 1993). Plasticity of central connections may be positive, that is, compensate the lack in target reinnervation, but may also result in maladaptive changes, such as neuropathic pain, hyperreflexia, dystonia, and phantom limb awareness (Navarro et al., 2007).

The response of the CNS to altered peripheral inputs may take many forms and include changes in ongoing or stimulus-evoked activity, neurochemical changes, functional alterations of excitatory and inhibitory connections, atrophy and degeneration of normal substrates, sprouting of new connections, and reorganization of somatosensory and motor maps (Davis et al., 2011; Navarro et al., 2007).

Plasticity of the somatosensory system has been extensively studied, and it has been shown that dramatic changes in the organization of cortical topography of the S1 area (primary somatosensory cortex) occur in response to a peripheral nerve injury (Donoghue & Sanes, 1987). It has been demonstrated that following peripheral nerve lesion in adult monkeys, the area in the somatosensory cortex corresponding to the deafferented body parts of the S1 became responsive to inputs from neighboring body parts (Merzenich et al., 1983). Although this form of cortical plasticity is well documented across several sensory systems and in several species, such as cats, raccoons, rodents, and bats, the understanding of the underlying mechanisms remains an active area of research (Pelled, Chuang, Dodd, & Koretsky, 2007).

In the motor system, changes in cortical representation also occur after peripheral injury; following amputation or peripheral nerve lesions, the area from which stimulation evoked movements of the adjacent body parts enlarged and the threshold for eliciting these movements is reduced (Donoghue & Sanes, 1988; Sanes, Suner, & Donoghue, 1990).

The hypothesis that also visceral afferents exert a significant influence on the CNS plasticity has been investigated by many researchers, even if little is known. Recent studies have showed that vagus nerve stimulation (VNS, a well-established adjunctive treatment for intractable epilepsy and treatment-resistant depression) can induce neurogenesis and plasticity in the hippocampus (Biggio et al., 2009). In particular, it was demonstrated that acute VNS induces cell proliferation in the dentate gyrus of the adult rat hippocampus and an increase of both the total amount of doublecortin (DCX) immunoreactivity and the number of DCX-positive neurons in the dentate gyrus. In contrast, chronic VNS induced an increase of the BDNF expression, which may serve to promote and maintain new neuronal connections formed in response to chronic VNS (Biggio et al., 2009). Moreover, it was shown that acute VNS increased the expression of genes for BDNF and basic FGF in the rat hippocampus, both of which are important modulators of hippocampal plasticity and neurogenesis (Follesa et al., 2007). Finally, a recent study showed that damage to the subdiaphragmatic vagus in adult rats is followed by microglia activation and long-lasting changes in the dentate gyrus, leading to alteration of neurogenesis (Ronchi, Ryu, Fornaro, & Czaja, 2012).

5. HOW TO STUDY PERIPHERAL NERVE REGENERATION?

Essential progress in peripheral nerve regeneration research has been possible using animal models which may simulate anatomical, physiological, and behavioral aspects of the regenerative process. The experimental models available can be divided into three main groups according to (i) the animal model, (ii) the localization of lesion, and (iii) the type of lesion.

By far, in nerve regeneration studies, the most employed laboratory animals are rats. The main reason appears to be the larger physical size of rat nerves compared to mouse nerves, which reduces the complexity of the microsurgical procedures (Tos et al., 2008), the possibility to have standardized and comparable functional tests and the fact that rats are more resilient than mice. On the other hand, the availability of genetically modified mouse colonies will probably increase mouse employment since transgenic models will allow to elucidate the role of a particular gene or protein in the mechanisms of the nerve regeneration process. In addition to these models, various large animal models have been employed including rabbits, sheep, pigs, and primates because several authors believe that the translation to clinical application may benefit from a preclinical study on large animal nerves since the regeneration process of nerves in large animals is more similar to humans (Fullarton, Lenihan, Myles, & Glasby, 2000).

According to the localization of the lesion, until recently, most peripheral nerve regeneration studies had been mainly carried out on the rat sciatic nerve model, primarily because it is the largest peripheral nerve (Baptista et al., 2007; Luis et al., 2007; Varejao et al., 2004). However, since most of the human peripheral nerve injuries affect the upper extremity, the necessity of an experimental model closer to clinical interests is required. Indeed, recent years have shown an increasing interest toward the employment of major forelimb nerves (Geuna et al., 2007; Papalia, Tos, Scevola, Raimondo, & Geuna, 2006; Papalia, Tos, Stagno d'Alcontres, Battiston, & Geuna, 2003; Sinis et al., 2008; Tos et al., 2009). In particular, the median nerve attracted the attention of peripheral nerve researchers because of the availability of an easier and more reliable behavioral test (the grasping test) (Lee, Tos, et al., 2007; Papalia et al., 2003; Tos et al., 2007).

According to the type of lesion, so far, two main experimental lesion paradigms have been adopted for nerve regeneration studies: (1) axonotmesis (crush), which is characterized by complete interruption of nerve fibers continuity without discontinuing the nerve, and (2) neurotmesis, which is a complete transection of the whole nerve. The complete nerve transection requires surgical repair to reestablish epineurial continuity. This experimental paradigm provides not only the model for the comparative investigation of new types of microsurgical and tissue engineering approaches for nerve reconstruction but also a good model for assessing the effectiveness of various postoperative treatments (drugs, physical therapy, diet, etc.). On the other hand, with a crush lesion, the injured axons are provided with an optimal regeneration pathway, represented by the nerve segment distal to the injury, without the need for the microsurgical repair. This experimental approach is therefore less technically challenging and is particularly suitable when a reproducible regeneration process is required, such as for the study of the biological mechanisms of regeneration or rationale development for new therapeutic agents.

Recent advances in molecular neurobiology include the development of transgenic mice that have been used in multiple areas, including the field of developmental neurobiology and disease processes such as cancer and diabetes, but their use has extended also to the study of peripheral nerve regeneration subsequent to traumatic injuries.

In particular, knockout mice, which carry a targeted gene inactivated through genetic engineering, are useful to elucidate the role of a particular gene or protein in a physiologic pathway by evaluating the consequences of its inactivation. On the other hand, transgenic animals that overexpress a particular gene or proteins are available, thus circumventing methodological difficulties in drug delivery, maintenance of constant neurotrophic factor concentrations, and the comorbidities associated with achieving these aims. Moreover, the availability of conditional knockout mice whose mutations can be targeted both spatially and temporally obviate the problem that some homozygous knockout mice can be embryonically lethal, thus limiting their usefulness. In addition, emerging tools include mice the axons or Schwann cells of which express fluorescent chromophores, which enabled new experiments with direct visualization of nerve regeneration over time.

In the past years, many studies of peripheral nerve regeneration have used knockout animals to elucidate the role of different neurotrophic factors during the process. For example, it has been shown that animals lacking CNTF are unable to produce motor nerve terminal sprouts after nerve transection or botulinum toxin injection and have also decreased ability to repair peripheral nerve damage from crush injury (Mizisin, Vu, Shuff, & Calcutt, 2004). Studies using GDNF-deficient mice, as well as NT-3 deficient mice, revealed inadequate development of sympathetic and sensory neurons (Anand, 2004), whereas knockout animals for the low-affinity NGF receptor p75NGFR showed decreased sensory innervation (Lee et al., 1992). Transgenic mice lacking IGF-1 show a decrease in motor and sensory nerve conduction velocities but no significant reduction in peripheral nerve myelination (Gao et al., 1998).

Animals lacking ApoD revealed a decrease in motor nerve conduction velocity and thickness of myelin sheath in intact nerves, and after injury, axon regeneration and remyelination are delayed (Ganfornina et al., 2010). The lack of Cx32, a gap junction protein, showed abnormally thin myelin sheaths, reflecting myelin degeneration-induced Schwann cell proliferation, while nerve conductance properties are altered only slightly (Anzini et al., 1997). Neuropilin-2-deficient mice showed slower axonal regeneration, remyelination of the regenerating axons, and recovery of normal gait after a crush lesion of the sciatic nerve (Bannerman et al., 2008). An experiment using the Cre–loxP system to disrupt the laminin γ1gene in Schwann cells showed that, during development, Schwann cells that lack laminin γ1 were unable to differentiate and synthesize myelin proteins, and therefore unable to myelinate axons. Moreover, after sciatic nerve crush, the axons showed impaired regeneration in mutant mice (Chen & Strickland, 2003). Peripheral nerves develop and function normally in GFAP-null mice. However, axonal regeneration after crush lesion was delayed. Mutant Schwann cells maintained the ability to dedifferentiate but showed defective proliferation, a key event for successful nerve regeneration (Triolo et al., 2006).

Neurofilament light knockout mice develop normally, but the regeneration of myelinated axons following crush injury was found to be abnormal with fewer newly regenerated myelinated axons in the sciatic nerve and facial nerve (Zhu, Couillard-Despres, & Julien, 1997). BACE1 (β-site amyloid precursor protein cleaving enzyme 1) knockout and wild-type nerves degenerated at a similar rate after axotomy. However, BACE1 knockout mice had markedly enhanced clearance of axonal and myelin debris from degenerated fibers, accelerated axonal regeneration, and earlier reinnervation of NMJs, compared with controls (Farah et al., 2011).

Fricker et al. (2011) used a single-neuron labeling with inducible Cre-mediated knockout animals, which enabled visualization of a subset of adult myelinated sensory and motoneurons in which Nrg1 was inducibly mutated by tamoxifen treatment. In uninjured mice, NRG1-deficient axons and the associated myelin sheath were normal, and the NMJ demonstrated normal apposition of presynaptic and postsynaptic components. After sciatic nerve crush, NRG1 ablation resulted in severe defects in remyelination: axons were either hypomyelinated or had no myelin sheath. NRG1-deficient axons were also found to regenerate at a slower rate (Fricker et al., 2011).

Other studies use the overexpression of proteins in order to better understand the role of a certain protein and to circumvent methodological difficulties in drug delivery. For example, transgenic mice constitutively expressing both interleukin 6 (IL-6) and its receptor (IL-6R) showed accelerated regeneration of the axotomized nerve (Hirota, Kiyama, Kishimoto, & Taga, 1996). The overexpression of FGF-2 showed no difference in number and size of myelinated fibers compared to wild-type mice in intact nerves. On the other hand, 1 week after crush injury, the number of regenerated axons was doubled and the myelin thickness was significantly smaller in transgenic mice, but after 2 and 4 weeks, there were no differences in the recovery of sensory and motor nerve fibers, showing that FGF-2 influences early peripheral nerve regeneration by regulating Schwann cell proliferation, axonal regrowth, and remyelination (Jungnickel, Haase, Konitzer, Timmer, & Grothe, 2006). Transgenic mice expressing Nogo-C in peripheral Schwann cells regenerate axons less rapidly than do wild-type mice after mid-thigh sciatic nerve crush (Kim, Bonilla, Qiu, & Strittmatter, 2003). On the other hand, using regulated transgenic expression of Nogo-A in peripheral nerve Schwann cells, Pot et al. (2002) showed that axonal regeneration and functional recovery are impaired after a sciatic nerve crush. Finally, the overexpression of L1(adhesion molecule) in neurons had no effect on femoral nerve function, numbers of quadriceps motoneurons, and myelinated axons in injured nerves; after femoral nerve injury, L1 overexpression had no impact on the time course and degree of functional recovery, but myelination in the motor and sensory femoral nerve branches was significantly improved and loss of perisomatic inhibitory terminals on motoneurons was attenuated in the transgenic mice (Guseva et al., 2011). Recently, we showed that constitutive ErbB2 receptor overexpression improves nerve regeneration following traumatic injury, possibly through the upregulation of soluble NRG1 isoforms (Ronchi et al., 2013).

6. CONCLUSIONS

Recent advances in peripheral nerve regeneration research have strengthened the view that the process of nerve regeneration and target reinnervation is quite complex and involves several factors. Keeping in mind the complexity of the whole process is thus very important for improving our knowledge on peripheral nerve regeneration, especially in the perspective of translating basic science results to the clinics. Figure 8.2 synthesizes this concept: in fact, it is important, but not enough, to study the involvement of a

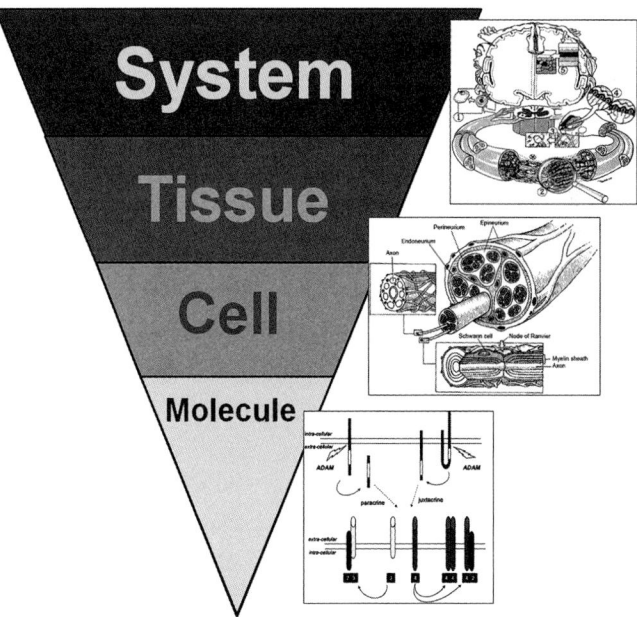

Figure 8.2 Scheme representing the complexity of the regenerative process that occur after a peripheral nerve injury. (For color version of this figure, the reader is referred to the online version of this chapter.)

single molecule, such as the NRG1/ErbB system (Ronchi et al., 2013) or other ligands and receptors involving in the process. It is also important, but not enough, to study the cell–cell interaction (in particular the interaction between regenerating axons and glial cells). Moreover, it is important, but not enough, to study the dynamics of the regenerative nerve, for example, with a tissue engineering approach and reconstructive microsurgery.

Rather, the whole system must be taken into account, including not only the damaged nerve and all the molecules involved in the process but also the proximal plasticity that occurs after a peripheral injury (DRG and CNS) and the effects that an injury has on distal sites (such as skeletal muscle atrophy).

Methods to improve the regenerative process should therefore simultaneously potentiate axonal regeneration, increase neuronal survival, and modulate central reorganization, as well as reduce muscle and sensory receptor atrophy and degeneration. It is expected that this holistic approach might lead to significant improvement in the functional outcome and thus the quality of life of the patients suffering from peripheral nerve injury.

ACKNOWLEDGMENTS

The research leading to this chapter has received funding from the European Community's Seventh Framework Programme (FP7-HEALTH-2011) under grant agreement no. 278612 (BIOHYBRID), from MIUR, and from Compagnia di San Paolo (MOVAG).

REFERENCES

Anand, P. (2004). Neurotrophic factors and their receptors in human sensory neuropathies. *Progress in Brain Research, 146,* 477–492.

Anzini, P., Neuberg, D. H., Schachner, M., Nelles, E., Willecke, K., Zielasek, J., et al. (1997). Structural abnormalities and deficient maintenance of peripheral nerve myelin in mice lacking the gap junction protein connexin 32. *The Journal of Neuroscience, 17*(12), 4545–4551.

Bain, J. R., Veltri, K. L., Chamberlain, D., & Fahnestock, M. (2001). Improved functional recovery of denervated skeletal muscle after temporary sensory nerve innervation. *Neuroscience, 103*(2), 503–510.

Bannerman, P., Ara, J., Hahn, A., Hong, L., McCauley, E., Friesen, K., et al. (2008). Peripheral nerve regeneration is delayed in neuropilin 2-deficient mice. *Journal of Neuroscience Research, 86*(14), 3163–3169.

Baptista, A. F., Gomes, J. R., Oliveira, J. T., Santos, S. M., Vannier-Santos, M. A., & Martinez, A. M. (2007). A new approach to assess function after sciatic nerve lesion in the mouse—Adaptation of the sciatic static index. *Journal of Neuroscience Methods, 161*(2), 259–264.

Battiston, B., Geuna, S., Ferrero, M., & Tos, P. (2005). Nerve repair by means of tubulization: Literature review and personal clinical experience comparing biological and synthetic conduits for sensory nerve repair. *Microsurgery, 25*(4), 258–267.

Battiston, B., Raimondo, S., Tos, P., Gaidano, V., Audisio, C., Scevola, A., et al. (2009). Chapter 11: Tissue engineering of peripheral nerves. *International Review of Neurobiology, 87,* 227–249.

Battiston, B., Tos, P., Cushway, T. R., & Geuna, S. (2000). Nerve repair by means of vein filled with muscle grafts. I. Clinical results. *Microsurgery*, *20*(1), 32–36.

Battiston, B., Tos, P., Geuna, S., Giacobini-Robecchi, M. G., & Guglielmone, R. (2000). Nerve repair by means of vein filled with muscle grafts. II. Morphological analysis of regeneration. *Microsurgery*, *20*(1), 37–41.

Biggio, F., Gorini, G., Utzeri, C., Olla, P., Marrosu, F., Mocchetti, I., et al. (2009). Chronic vagus nerve stimulation induces neuronal plasticity in the rat hippocampus. *The International Journal of Neuropsychopharmacology*, *12*(9), 1209–1221.

Birch, R., & Raji, A. R. (1991). Repair of median and ulnar nerves. Primary suture is best. *The Journal of Bone and Joint Surgery. British Volume*, *73*(1), 154–157.

Bruck, W. (1997). The role of macrophages in Wallerian degeneration. *Brain Pathology*, *7*(2), 741–752.

Brunelli, G. (2005). Direct muscular neurotization. *Journal of the American Society for Surgery of the Hand*, *5*(4), 193–200.

Brunelli, G. A., Battiston, B., Vigasio, A., Brunelli, G., & Marocolo, D. (1993). Bridging nerve defects with combined skeletal muscle and vein conduits. *Microsurgery*, *14*(4), 247–251.

Buchthal, F., & Kuhl, V. (1979). Nerve conduction, tactile sensibility, and the electromyogram after suture or compression of peripheral nerve: A longitudinal study in man. *Journal of Neurology, Neurosurgery, and Psychiatry*, *42*(5), 436–451.

Bungner, O. V. (1891). Ueber die degenerations und regenerations—Vorgange am Nerven nach Verletzungen. *Beitrage zur Pathologischen Anatomie und zur Allgemeinen Pathologie*, *10*, 321.

Burnett, M. G., & Zager, E. L. (2004). Pathophysiology of peripheral nerve injury: A brief review. *Neurosurgical Focus*, *16*(5), E1.

Carroll, S. L., Miller, M. L., Frohnert, P. W., Kim, S. S., & Corbett, J. A. (1997). Expression of neuregulins and their putative receptors, ErbB2 and ErbB3, is induced during Wallerian degeneration. *The Journal of Neuroscience*, *17*(5), 1642–1659.

Chen, Z. L., & Strickland, S. (2003). Laminin gamma1 is critical for Schwann cell differentiation, axon myelination, and regeneration in the peripheral nerve. *The Journal of Cell Biology*, *163*(4), 889.

Chen, Z. L., Yu, W. M., & Strickland, S. (2007). Peripheral regeneration. *Annual Review of Neuroscience*, *30*, 209–233.

Coggeshall, R. E., Lekan, H. A., Doubell, T. P., Allchorne, A., & Woolf, C. J. (1997). Central changes in primary afferent fibers following peripheral nerve lesions. *Neuroscience*, *77*(4), 1115–1122.

Cohen, J. A., Yachnis, A. T., Arai, M., Davis, J. G., & Scherer, S. S. (1992). Expression of the neu proto-oncogene by Schwann cells during peripheral nerve development and Wallerian degeneration. *Journal of Neuroscience Research*, *31*(4), 622–634.

Corti, S., Locatelli, F., Papadimitriou, D., Donadoni, C., Del Bo, R., Crimi, M., et al. (2006). Transplanted ALDHhiSSClo neural stem cells generate motor neurons and delay disease progression of nmd mice, an animal model of SMARD1. *Human Molecular Genetics*, *15*(2), 167–187.

Corti, S., Nizzardo, M., Nardini, M., Donadoni, C., Salani, S., Ronchi, D., et al. (2009). Embryonic stem cell-derived neural stem cells improve spinal muscular atrophy phenotype in mice. *Brain*, *133*(Pt. 2), 465–481.

Davis, K. D., Taylor, K. S., & Anastakis, D. J. (2011). Nerve injury triggers changes in the brain. *The Neuroscientist*, *17*(4), 407–422.

de Putter, C. E., Selles, R. W., Polinder, S., Panneman, M. J., Hovius, S. E., & van Beeck, E. F. (2012). Economic impact of hand and wrist injuries: Health-care costs and productivity costs in a population-based study. *The Journal of Bone and Joint Surgery. American Volume*, *94*(9), e56.

Deshpande, D. M., Kim, Y. S., Martinez, T., Carmen, J., Dike, S., Shats, I., et al. (2006). Recovery from paralysis in adult rats using embryonic stem cells. *Annals of Neurology*, *60*(1), 32–44.

Donoghue, J. P., & Sanes, J. N. (1987). Peripheral nerve injury in developing rats reorganizes representation pattern in motor cortex. *Proceedings of the National Academy of Sciences of the United States of America*, *84*(4), 1123–1126.

Donoghue, J. P., & Sanes, J. N. (1988). Organization of adult motor cortex representation patterns following neonatal forelimb nerve injury in rats. *The Journal of Neuroscience*, *8*(9), 3221–3232.

Dvali, L., & Mackinnon, S. (2003). Nerve repair, grafting, and nerve transfers. *Clinics in Plastic Surgery*, *30*(2), 203–221.

Farah, M. H., Pan, B. H., Hoffman, P. N., Ferraris, D., Tsukamoto, T., Nguyen, T., et al. (2011). Reduced BACE1 activity enhances clearance of myelin debris and regeneration of axons in the injured peripheral nervous system. *The Journal of Neuroscience*, *31*(15), 5744–5754.

Fawcett, J. W., & Keynes, R. J. (1990). Peripheral nerve regeneration. *Annual Review of Neuroscience*, *13*, 43–60.

Fenrich, K., & Gordon, T. (2004). Canadian Association of Neuroscience review: Axonal regeneration in the peripheral and central nervous systems—Current issues and advances. *The Canadian Journal of Neurological Sciences*, *31*(2), 142–156.

Fields, R. D., Le Beau, J. M., Longo, F. M., & Ellisman, M. H. (1989). Nerve regeneration through artificial tubular implants. *Progress in Neurobiology*, *33*(2), 87–134.

Finkelstein, D. I., Dooley, P. C., & Luff, A. R. (1993). Recovery of muscle after different periods of denervation and treatments. *Muscle & Nerve*, *16*(7), 769–777.

Follesa, P., Biggio, F., Gorini, G., Caria, S., Talani, G., Dazzi, L., et al. (2007). Vagus nerve stimulation increases norepinephrine concentration and the gene expression of BDNF and bFGF in the rat brain. *Brain Research*, *1179*, 28–34.

Fornaro, M., Tos, P., Geuna, S., Giacobini-Robecchi, M. G., & Battiston, B. (2001). Confocal imaging of Schwann-cell migration along muscle-vein combined grafts used to bridge nerve defects in the rat. *Microsurgery*, *21*(4), 153–155.

Fricker, F. R., Lago, N., Balarajah, S., Tsantoulas, C., Tanna, S., Zhu, N., et al. (2011). Axonally derived neuregulin-1 is required for remyelination and regeneration after nerve injury in adulthood. *The Journal of Neuroscience*, *31*(9), 3225–3233.

Fu, S. Y., & Gordon, T. (1995). Contributing factors to poor functional recovery after delayed nerve repair: Prolonged denervation. *The Journal of Neuroscience*, *15*(5 Pt. 2), 3886–3895.

Fullarton, A. C., Lenihan, D. V., Myles, L. M., & Glasby, M. A. (2000). Obstetric brachial plexus palsy: A large animal model for traction injury and its repair. Part 1: Age of the recipient. *Journal of Hand Surgery. British Volume*, *25*(1), 52–57.

Ganfornina, M. D., Do Carmo, S., Martinez, E., Tolivia, J., Navarro, A., Rassart, E., et al. (2010). ApoD, a glia-derived apolipoprotein, is required for peripheral nerve functional integrity and a timely response to injury. *Glia*, *58*(11), 1320–1334.

Gao, J., Coggeshall, R. E., Tarasenko, Y. I., & Wu, P. (2005). Human neural stem cell-derived cholinergic neurons innervate muscle in motoneuron deficient adult rats. *Neuroscience*, *131*(2), 257–262.

Gao, W. Q., Shinsky, N., Armanini, M. P., Moran, P., Zheng, J. L., Mendoza-Ramirez, J. L., et al. (1998). Regulation of hippocampal synaptic plasticity by the tyrosine kinase receptor, REK7/EphA5, and its ligand, AL-1/Ephrin-A5. *Molecular and Cellular Neuroscience*, *11*(5–6), 247–259.

Geuna, S., Raimondo, S., Ronchi, G., Di Scipio, F., Tos, P., Czaja, K., et al. (2009). Chapter 3: Histology of the peripheral nerve and changes occurring during nerve regeneration. *International Review of Neurobiology*, *87*, 27–46.

Geuna, S., Tos, P., Raimondo, S., Lee, J. M., Gambarotta, G., Nicolino, S., et al. (2007). Functional, morphological and biomolecular assessment of posttraumatic neuromuscular recovery in the rat forelimb model. *Acta Neurochirurgica. Supplement, 100,* 173–177.

Glasby, M. A., Gschmeissner, S., Hitchcock, R. J., & Huang, C. L. (1986). Regeneration of the sciatic nerve in rats. The effect of muscle basement membrane. *The Journal of Bone and Joint Surgery. British Volume, 68*(5), 829–833.

Griffin, J. W., & Thompson, W. J. (2008). Biology and pathology of nonmyelinating Schwann cells. *Glia, 56*(14), 1518–1531.

Guseva, D., Zerwas, M., Xiao, M. F., Jakovcevski, I., Irintchev, A., & Schachner, M. (2011). Adhesion molecule L1 overexpressed under the control of the neuronal Thy-1 promoter improves myelination after peripheral nerve injury in adult mice. *Experimental Neurology, 229*(2), 339–352.

Gutmann, E. (1962). Denervation and disuse atrophy in crosstriated muscle. *Revue Canadienne de Biologie, 21,* 353–365.

Haas, C. A., Donath, C., & Kreutzberg, G. W. (1993). Differential expression of immediate early genes after transection of the facial nerve. *Neuroscience, 53*(1), 91–99.

Hall, S. (2001). Nerve repair: A neurobiologist's view. *Journal of Hand Surgery. British Volume, 26*(2), 129–136.

Himes, B. T., & Tessler, A. (1989). Death of some dorsal root ganglion neurons and plasticity of others following sciatic nerve section in adult and neonatal rats. *The Journal of Comparative Neurology, 284*(2), 215–230.

Hirota, H., Kiyama, H., Kishimoto, T., & Taga, T. (1996). Accelerated nerve regeneration in mice by upregulated expression of interleukin (IL) 6 and IL-6 receptor after trauma. *The Journal of Experimental Medicine, 183*(6), 2627–2634.

Hopkins, W. G., & Slack, J. R. (1981). The sequential development of nodal sprouts in mouse muscles in response to nerve degeneration. *Journal of Neurocytology, 10*(4), 537–556.

Huang, J. K., Phillips, G. R., Roth, A. D., Pedraza, L., Shan, W., Belkaid, W., et al. (2005). Glial membranes at the node of Ranvier prevent neurite outgrowth. *Science, 310*(5755), 1813–1817.

Huelsenbeck, S. C., Rohrbeck, A., Handreck, A., Hellmich, G., Kiaei, E., Roettinger, I., et al. (2012). C3 peptide promotes axonal regeneration and functional motor recovery after peripheral nerve injury. *Neurotherapeutics, 9*(1), 185–198.

Hynes, N. M., Bain, J. R., Thoma, A., Veltri, K., & Maguire, J. A. (1997). Preservation of denervated muscle by sensory protection in rats. *Journal of Reconstructive Microsurgery, 13*(5), 337–343.

Ide, C. (1996). Peripheral nerve regeneration. *Neuroscience Research, 25*(2), 101–121.

Irintchev, A., Draguhn, A., & Wernig, A. (1990). Reinnervation and recovery of mouse soleus muscle after long-term denervation. *Neuroscience, 39*(1), 231–243.

Jain, N., Florence, S. L., & Kaas, J. H. (1998). Reorganization of somatosensory cortex after nerve and spinal cord injury. *News in Physiological Sciences, 13,* 143–149.

Jessen, K. R., & Mirsky, R. (2008). Negative regulation of myelination: Relevance for development, injury, and demyelinating disease. *Glia, 56*(14), 1552–1565.

Jungnickel, J., Haase, K., Konitzer, J., Timmer, M., & Grothe, C. (2006). Faster nerve regeneration after sciatic nerve injury in mice over-expressing basic fibroblast growth factor. *Journal of Neurobiology, 66*(9), 940–948.

Kaas, J. H. (1991). Plasticity of sensory and motor maps in adult mammals. *Annual Review of Neuroscience, 14,* 137–167.

Kaas, J. H., & Collins, C. E. (2003). Anatomic and functional reorganization of somatosensory cortex in mature primates after peripheral nerve and spinal cord injury. *Advances in Neurology, 93,* 87–95.

Kandel, E. R., & Squire, L. R. (2000). Neuroscience: Breaking down scientific barriers to the study of brain and mind. *Science, 290*(5494), 1113–1120.

Keynes, R. J., Hopkins, W. G., & Huang, L. H. (1984). Regeneration of mouse peripheral nerves in degenerating skeletal muscle: Guidance by residual muscle fibre basement membrane. *Brain Research, 295*(2), 275–281.

Kim, J. E., Bonilla, I. E., Qiu, D., & Strittmatter, S. M. (2003). Nogo-C is sufficient to delay nerve regeneration. *Molecular and Cellular Neuroscience, 23*(3), 451–459.

Latres, E., Amini, A. R., Amini, A. A., Griffiths, J., Martin, F. J., Wei, Y., et al. (2005). Insulin-like growth factor-1 (IGF-1) inversely regulates atrophy-induced genes via the phosphatidylinositol 3-kinase/Akt/mammalian target of rapamycin (PI3K/Akt/mTOR) pathway. *The Journal of Biological Chemistry, 280*(4), 2737–2744.

Leach, J. B., & Schmidt, C. E. (2005). Characterization of protein release from photocrosslinkable hyaluronic acid-polyethylene glycol hydrogel tissue engineering scaffolds. *Biomaterials, 26*(2), 125–135.

Lee, J. P., Jeyakumar, M., Gonzalez, R., Takahashi, H., Lee, P. J., Baek, R. C., et al. (2007). Stem cells act through multiple mechanisms to benefit mice with neurodegenerative metabolic disease. *Nature Medicine, 13*(4), 439–447.

Lee, K. F., Li, E., Huber, L. J., Landis, S. C., Sharpe, A. H., Chao, M. V., et al. (1992). Targeted mutation of the gene encoding the low affinity NGF receptor p75 leads to deficits in the peripheral sensory nervous system. *Cell, 69*(5), 737–749.

Lee, H. K., Shin, Y. K., Jung, J., Seo, S. Y., Baek, S. Y., & Park, H. T. (2009). Proteasome inhibition suppresses Schwann cell dedifferentiation in vitro and in vivo. *Glia, 57*(16), 1825–1834.

Lee, J. M., Tos, P., Raimondo, S., Fornaro, M., Papalia, I., Geuna, S., et al. (2007). Lack of topographic specificity in nerve fiber regeneration of rat forelimb mixed nerves. *Neuroscience, 144*(3), 985–990.

Letourneau, P. C., Condic, M. L., & Snow, D. M. (1994). Interactions of developing neurons with the extracellular matrix. *The Journal of Neuroscience, 14*(3 Pt. 1), 915–928.

Lewin, G. R., & McMahon, S. B. (1993). Muscle afferents innervating skin form somatotopically appropriate connections in the adult rat dorsal horn. *The European Journal of Neuroscience, 5*(8), 1083–1092.

Lieberman, A. R. (1971). The axon reaction: A review of the principal features of perikaryal responses to axon injury. *International Review of Neurobiology, 14*, 49–124.

Lu, D. X., Huang, S. K., & Carlson, B. M. (1997). Electron microscopic study of long-term denervated rat skeletal muscle. *The Anatomical Record, 248*(3), 355–365.

Lubinska, L. (1982). Patterns of Wallerian degeneration of myelinated fibres in short and long peripheral stumps and in isolated segments of rat phrenic nerve. Interpretation of the role of axoplasmic flow of the trophic factor. *Brain Research, 233*(2), 227–240.

Luis, A. L., Amado, S., Geuna, S., Rodrigues, J. M., Simoes, M. J., Santos, J. D., et al. (2007). Long-term functional and morphological assessment of a standardized rat sciatic nerve crush injury with a non-serrated clamp. *Journal of Neuroscience Methods, 163*(1), 92–104.

Lundborg, G. (2000). A 25-year perspective of peripheral nerve surgery: Evolving neuroscientific concepts and clinical significance. *Journal of Hand Surgery. American Volume, 25*(3), 391–414.

Lundborg, G. (2003). Richard P. Bunge memorial lecture. Nerve injury and repair—A challenge to the plastic brain. *Journal of the Peripheral Nervous System, 8*(4), 209–226.

Lundborg, G. (2004). Alternatives to autologous nerve grafts. *Handchirurgie, Mikrochirurgie, Plastische Chirurgie, 36*(1), 1–7.

Lundborg, G. (2005). *Nerve injury and repair: Regeneration, reconstruction, and cortical remodeling* (2nd ed.). Philadelphia: Churchill Livingstone.

Mackinnon, S. E. (1989). New directions in peripheral nerve surgery. *Annals of Plastic Surgery, 22*(3), 257–273.

McKay Hart, A., Brannstrom, T., Wiberg, M., & Terenghi, G. (2002). Primary sensory neurons and satellite cells after peripheral axotomy in the adult rat: Timecourse of cell death and elimination. *Experimental Brain Research, 142*(3), 308–318.

McQuarrie, I. G. (1985). Effect of conditioning lesion on axonal sprout formation at nodes of Ranvier. *The Journal of Comparative Neurology, 231*(2), 239–249.

Merle, M., Bour, C., Foucher, G., & Saint Laurent, Y. (1986). Sarcoid tenosynovitis in the hand. A case report and literature review. *Journal of Hand Surgery. British Volume, 11*(2), 281–286.

Merzenich, M. M., Kaas, J. H., Wall, J., Nelson, R. J., Sur, M., & Felleman, D. (1983). Topographic reorganization of somatosensory cortical areas 3b and 1 in adult monkeys following restricted deafferentation. *Neuroscience, 8*(1), 33–55.

Millesi, H. (1970). Problems in the reconstruction of extensive soft tissue defects in the upper extremeties. *Handchirurgie, 2*(2), 80–87.

Mizisin, A. P., Vu, Y., Shuff, M., & Calcutt, N. A. (2004). Ciliary neurotrophic factor improves nerve conduction and ameliorates regeneration deficits in diabetic rats. *Diabetes, 53*(7), 1807–1812.

Navarro, X., Rodriguez, F. J., Labrador, R. O., Buti, M., Ceballos, D., Gomez, N., et al. (1996). Peripheral nerve regeneration through bioresorbable and durable nerve guides. *Journal of the Peripheral Nervous System, 1*(1), 53–64.

Navarro, X., Vivo, M., & Valero-Cabre, A. (2007). Neural plasticity after peripheral nerve injury and regeneration. *Progress in Neurobiology, 82*(4), 163–201.

Ng, Y. P., Pun, S., Yang, J. F., Ip, N. Y., & Tsim, K. W. (1997). Chick muscle expresses various ARIA isoforms: Regulation during development, denervation, and regeneration. *Molecular and Cellular Neuroscience, 9*(2), 132–143.

Nicolaidis, S. C., & Williams, H. B. (2001). Muscle preservation using an implantable electrical system after nerve injury and repair. *Microsurgery, 21*(6), 241–247.

Nicolino, S., Panetto, A., Raimondo, S., Gambarotta, G., Guzzini, M., Fornaro, M., et al. (2009). Denervation and reinnervation of adult skeletal muscle modulate mRNA expression of neuregulin-1 and ErbB receptors. *Microsurgery, 29*(6), 464–472.

Papalia, I., Geuna, S., Tos, P. L., Boux, E., Battiston, B., & Stagno D'Alcontres, F. (2003). Morphologic and functional study of rat median nerve repair by terminolateral neurorrhaphy of the ulnar nerve. *Journal of Reconstructive Microsurgery, 19*(4), 257–264.

Papalia, I., Tos, P., Scevola, A., Raimondo, S., & Geuna, S. (2006). The ulnar test: A method for the quantitative functional assessment of posttraumatic ulnar nerve recovery in the rat. *Journal of Neuroscience Methods, 154*(1–2), 198–203.

Papalia, I., Tos, P., Stagno d'Alcontres, F., Battiston, B., & Geuna, S. (2003). On the use of the grasping test in the rat median nerve model: A re-appraisal of its efficacy for quantitative assessment of motor function recovery. *Journal of Neuroscience Methods, 127*(1), 43–47.

Pelled, G., Chuang, K. H., Dodd, S. J., & Koretsky, A. P. (2007). Functional MRI detection of bilateral cortical reorganization in the rodent brain following peripheral nerve deafferentation. *NeuroImage, 37*(1), 262–273.

Pfister, B. J., Gordon, T., Loverde, J. R., Kochar, A. S., Mackinnon, S. E., & Cullen, D. K. (2011). Biomedical engineering strategies for peripheral nerve repair: Surgical applications, state of the art, and future challenges. *Critical Reviews in Biomedical Engineering, 39*(2), 81–124.

Pfister, L. A., Papaloizos, M., Merkle, H. P., & Gander, B. (2007). Nerve conduits and growth factor delivery in peripheral nerve repair. *Journal of the Peripheral Nervous System, 12*(2), 65–82.

Porporato, P. E., Filigheddu, N., Reano, S., Ferrara, M., Angelino, E., Gnocchi, V. F., et al. (2013). Acylated and unacylated ghrelin impair skeletal muscle atrophy in mice. *The Journal of Clinical Investigation, 123*(2), 611–622.

Pot, C., Simonen, M., Weinmann, O., Schnell, L., Christ, F., Stoeckle, S., et al. (2002). Nogo-A expressed in Schwann cells impairs axonal regeneration after peripheral nerve injury. *The Journal of Cell Biology, 159*(1), 29–35.

Ronchi, G., Gambarotta, G., Di Scipio, F., Salamone, P., Sprio, A. E., Cavallo, F., et al. (2013). ErbB2 receptor over-expression improves post-traumatic peripheral nerve regeneration in adult mice. *PLoS One, 8*(2), e56282.

Ronchi, G., Ryu, V., Fornaro, M., & Czaja, K. (2012). Hippocampal plasticity after a vagus nerve injury in the rat. *Neural Regeneration Research, 7*(14), 1055–1063.

Sanes, J. N., Suner, S., & Donoghue, J. P. (1990). Dynamic organization of primary motor cortex output to target muscles in adult rats. I. Long-term patterns of reorganization following motor or mixed peripheral nerve lesions. *Experimental Brain Research, 79*(3), 479–491.

Schlaepfer, W. W. (1977). Structural alterations of peripheral nerve induced by the calcium ionophore A23187. *Brain Research, 136*(1), 1–9.

Schmalbruch, H., al-Amood, W. S., & Lewis, D. M. (1991). Morphology of long-term denervated rat soleus muscle and the effect of chronic electrical stimulation. *The Journal of Physiology, 441*, 233–241.

Schmidt, C. E., & Leach, J. B. (2003). Neural tissue engineering: Strategies for repair and regeneration. *Annual Review of Biomedical Engineering, 5*, 293–347.

Scholz, T., Krichevsky, A., Sumarto, A., Jaffurs, D., Wirth, G. A., Paydar, K., et al. (2009). Peripheral nerve injuries: An international survey of current treatments and future perspectives. *Journal of Reconstructive Microsurgery, 25*(6), 339–344.

Schultz, E., Jaryszak, D. L., & Valliere, C. R. (1985). Response of satellite cells to focal skeletal muscle injury. *Muscle & Nerve, 8*(3), 217–222.

Seddon, H. (1943). Three types of nerve injury. *Brain, 66*, 237–288.

Seddon, B., & Fynn, G. H. (1972). Cation activated hydrolysis of ATP by the soluble fraction of Bacillus brevis. *Microbios, 6*(22), 87–96.

Shen, Z. L., Lassner, F., Bader, A., Becker, M., Walter, G. F., & Berger, A. (2000). Cellular activity of resident macrophages during Wallerian degeneration. *Microsurgery, 20*(5), 255–261.

Sinis, N., Guntinas-Lichius, O., Irintchev, A., Skouras, E., Kuerten, S., Pavlov, S. P., et al. (2008). Manual stimulation of forearm muscles does not improve recovery of motor function after injury to a mixed peripheral nerve. *Experimental Brain Research, 185*(3), 469–483.

Stoll, G., Griffin, J. W., Li, C. Y., & Trapp, B. D. (1989). Wallerian degeneration in the peripheral nervous system: Participation of both Schwann cells and macrophages in myelin degradation. *Journal of Neurocytology, 18*(5), 671–683.

Stoll, G., Jander, S., & Myers, R. R. (2002). Degeneration and regeneration of the peripheral nervous system: From Augustus Waller's observations to neuroinflammation. *Journal of the Peripheral Nervous System, 7*(1), 13–27.

Suarez, E., Bach, D., Cadefau, J., Palacin, M., Zorzano, A., & Guma, A. (2001). A novel role of neuregulin in skeletal muscle. Neuregulin stimulates glucose uptake, glucose transporter translocation, and transporter expression in muscle cells. *The Journal of Biological Chemistry, 276*(21), 18257–18264.

Sun, W., Sun, C., Zhao, H., Lin, H., Han, Q., Wang, J., et al. (2009). Improvement of sciatic nerve regeneration using laminin-binding human NGF-beta. *PLoS One, 4*(7), e6180.

Sunderland, S. (1951). A classification of peripheral nerve injuries producing loss of function. *Brain, 74*(4), 491–516.

Swett, J. E., Hong, C. Z., & Miller, P. G. (1995). Most dorsal root ganglion neurons of the adult rat survive nerve crush injury. *Somatosensory and Motor Research, 12*(3–4), 177–189.

Tang, H., Cheung, W. M., Ip, F. C., & Ip, N. Y. (2000). Identification and characterization of differentially expressed genes in denervated muscle. *Molecular and Cellular Neuroscience, 16*(2), 127–140.

Terenghi, G. (1999). Peripheral nerve regeneration and neurotrophic factors. *Journal of Anatomy, 194*(Pt. 1), 1–14.

Terzis, J. K., Sun, D. D., & Thanos, P. K. (1997). Historical and basic science review: Past, present, and future of nerve repair. *Journal of Reconstructive Microsurgery, 13*(3), 215–225.

Tetzlaff, W., Gilad, V. H., Leonard, C., Bisby, M. A., & Gilad, G. M. (1988). Retrograde changes in transglutaminase activity after peripheral nerve injuries. *Brain Research, 445*(1), 142–146.

Tos, P., Battiston, B., Ciclamini, D., Geuna, S., & Artiaco, S. (2012). Primary repair of crush nerve injuries by means of biological tubulization with muscle-vein-combined grafts. *Microsurgery, 32*(5), 358–363.

Tos, P., Battiston, B., Nicolino, S., Raimondo, S., Fornaro, M., Lee, J. M., et al. (2007). Comparison of fresh and predegenerated muscle-vein-combined guides for the repair of rat median nerve. *Microsurgery, 27*(1), 48–55.

Tos, P., Ronchi, G., Nicolino, S., Audisio, C., Raimondo, S., Fornaro, M., et al. (2008). Employment of the mouse median nerve model for the experimental assessment of peripheral nerve regeneration. *Journal of Neuroscience Methods, 169*(1), 119–127.

Tos, P., Ronchi, G., Papalia, I., Sallen, V., Legagneux, J., Geuna, S., et al. (2009). Chapter 4: Methods and protocols in peripheral nerve regeneration experimental research: Part I— Experimental models. *International Review of Neurobiology, 87*, 47–79.

Trapp, B. D., Hauer, P., & Lemke, G. (1988). Axonal regulation of myelin protein mRNA levels in actively myelinating Schwann cells. *The Journal of Neuroscience, 8*(9), 3515–3521.

Triolo, D., Dina, G., Lorenzetti, I., Malaguti, M., Morana, P., Del Carro, U., et al. (2006). Loss of glial fibrillary acidic protein (GFAP) impairs Schwann cell proliferation and delays nerve regeneration after damage. *Journal of Cell Science, 119*(Pt. 19), 3981–3993.

Varejao, A. S., Cabrita, A. M., Meek, M. F., Bulas-Cruz, J., Melo-Pinto, P., Raimondo, S., et al. (2004). Functional and morphological assessment of a standardized rat sciatic nerve crush injury with a non-serrated clamp. *Journal of Neurotrauma, 21*(11), 1652–1670.

Vargas, M. E., & Barres, B. A. (2007). Why is Wallerian degeneration in the CNS so slow? *Annual Review of Neuroscience, 30*, 153–179.

Veltri, K., Kwiecien, J. M., Minet, W., Fahnestock, M., & Bain, J. R. (2005). Contribution of the distal nerve sheath to nerve and muscle preservation following denervation and sensory protection. *Journal of Reconstructive Microsurgery, 21*(1), 57–70, discussion 71–54.

Vestergaard, S., Tandrup, T., & Jakobsen, J. (1997). Effect of permanent axotomy on number and volume of dorsal root ganglion cell bodies. *The Journal of Comparative Neurology, 388*(2), 307–312.

Vial, J. D. (1958). The early changes in the axoplasm during Wallerian degeneration. *The Journal of Biophysical and Biochemical Cytology, 4*(5), 551–555.

Wall, J. T., Xu, J., & Wang, X. (2002). Human brain plasticity: An emerging view of the multiple substrates and mechanisms that cause cortical changes and related sensory dysfunctions after injuries of sensory inputs from the body. *Brain Research. Brain Research Reviews, 39*(2–3), 181–215.

Wang, H., Gu, Y., Xu, J., Shen, L., & Li, J. (2001). Comparative study of different surgical procedures using sensory nerves or neurons for delaying atrophy of denervated skeletal muscle. *Journal of Hand Surgery. American Volume, 26*(2), 326–331.

White, F. V., Toews, A. D., Goodrum, J. F., Novicki, D. L., Bouldin, T. W., & Morell, P. (1989). Lipid metabolism during early stages of Wallerian degeneration in the rat sciatic nerve. *Journal of Neurochemistry, 52*(4), 1085–1092.

Wrede, L. (1909). Uberbrueckung eines Nervendefektes mittels Seidennahtund leben Venenstueckes. *Deutsche Medizinische Wochenschrift, 35*, 1125–1160.

Yamaguchi, A., Ishii, H., Morita, I., Oota, I., & Takeda, H. (2004). mRNA expression of fibroblast growth factors and hepatocyte growth factor in rat plantaris muscle following denervation and compensatory overload. *Pflügers Archiv, 448*(5), 539–546.

Yannas, I. V., & Hill, B. J. (2004). Selection of biomaterials for peripheral nerve regeneration using data from the nerve chamber model. *Biomaterials, 25*(9), 1593–1600.

Yoshida, T., Semprun-Prieto, L., Sukhanov, S., & Delafontaine, P. (2010). IGF-1 prevents ANG II-induced skeletal muscle atrophy via Akt- and Foxo-dependent inhibition of the ubiquitin ligase atrogin-1 expression. *American Journal of Physiology. Heart and Circulatory Physiology, 298*(5), H1565–H1570.

Young, R. C., Wiberg, M., & Terenghi, G. (2002). Poly-3-hydroxybutyrate (PHB): A resorbable conduit for long-gap repair in peripheral nerves. *British Journal of Plastic Surgery, 55*(3), 235–240.

Zhao, C., Veltri, K., Li, S., Bain, J. R., & Fahnestock, M. (2004). NGF, BDNF, NT-3, and GDNF mRNA expression in rat skeletal muscle following denervation and sensory protection. *Journal of Neurotrauma, 21*(10), 1468–1478.

Zhu, Q., Couillard-Despres, S., & Julien, J. P. (1997). Delayed maturation of regenerating myelinated axons in mice lacking neurofilaments. *Experimental Neurology, 148*(1), 299–316.

INDEX

Note: Page numbers followed by "*f*" indicate figures, and "*t*" indicate tables.

A

Acetylcholine receptor (AChR)
  intact gastrocnemius muscle, rats, 102, 103*f*
  internal and membrane inserted, 101
  muscle denervation, 100
AChR. *See* Acetylcholine receptor (AChR)
American Society of Testing Materials (ASTM F04 division IV), 4
Apoptosis, 156–157
Autologous nerve grafts, 172
Axonal regeneration, 91–92, 92*f*, 93, 95*f*, 96
Axonotmesis, 168–169, 180

B

BDNF. *See* Brain-derived neurotrophic factor (BDNF)
Biocompatibility, chitosan, 4–5, 9–11
Biodegradability, chitosan, 4–5, 7–9
Biological effects, peripheral nerve regeneration
  axonal regeneration, 117–118
  neurotrophic factors, 118–119
  Schwann cells, 119–120
Biomedical applications, nerve interfaces
  application modes, peripheral systems, 73, 73*f*
  CNS-injured patients, 73–76
  prostheses control, 76–77
Brain-derived neurotrophic factor (BDNF), 128–134, 135–137, 138–139, 140–141
Brain plasticity. *See* Cerebral plasticity

C

Carpal tunnel syndrome, 141, 142
CCI. *See* Chronic constriction injury (CCI)
Central nervous system (CNS) repair
  autograft implantation, 16
  biomaterials, 15–16
  degradation kinetics, 15–16
  ex vivo and *in vivo* SCI models, 16
  neurotrophic factors/neuroprotective molecules, 22*t*, 25

SCI, 14–16
supportive cells, 19*t*, 24–25
surface modification, 16–24, 18*t*
Cerebral plasticity
  diagnosis, 161
  functional recovery, 160–161
  nerve and repair and reconstruction, 160–161
  rehabilitation strategies, 160–161
  "Sensor Glove System", 160–161
  timing, 160–161
  younger children, 160–161
Chitin
  biocompatibility, 9
  biodegradability, 7–8
  α-chitin, 2
  chitosan (*see* Chitosan-based scaffold)
  deacetylation, 3
  description, 2, 3
  extraction, 2
  solubilities, 3
Chitosan-based scaffold
  CNS repair, 14–25
  description, 2
  drug delivery systems, 5
  as "fusogen", 4
  in vitro evidence, 5–14
  limitations, applications, 3
  medical and pharmaceutical applications, 3
  peripheral nervous system repair, 25–48
  preparation, 3
  saturated steam and γ irradiation, 4
  SurgiLux, 3
Chronic constriction injury (CCI), 140–141
CK. *See* Creatine kinase (CK)
CNS-injured patients
  foot drop correction, 74–75
  hand grasping prostheses, 75–76
  sacral root stimulation, 74
  ventilatory pacing, 74
  walking assistance prostheses, 75

CONTENTS OF RECENT VOLUMES

Volume 42

Volume 78

Volume 79

Volume 82

Volume 85

Volume 86

Volume 87

Volume 92

Volume 93